# The 1914–2024 War Atlas

*The 1914–2024 War Atlas* deconstructs the contemporary widespread and well-known image of the 20th and 21st centuries, arguing for the continuity of the historical process covering the period 1914–2024.

The years between 1914 and 2024 constitute a period of unparalleled economic growth, scientific advancement, political development, and social change – in just over 100 years, human civilization has moved from an industrial to a digital age. However, they also cover some of the most dangerous, violent, and politically volatile years in human history. In these years, two world wars have been fought; weapons of mass destruction dropped; authoritarian and democratic regimes alike have risen and fallen; and regional conflicts have been almost continuous, effectively conjoining World War I, World War II, the Cold War, and the Russo-Ukrainian war. This volume indicates that the 1914 July Crisis set in motion a sequence of events that spanned over 100 years. Containing a range of colourful maps and charts, this book graphically illustrates the arguments presented in both an informative and visual way.

This atlas will serve as a perfect textbook for students studying history, geography, politics, and international relations, as well as being a useful guide to contemporary world politics for researchers and for those interested in international relations and modern history.

**Marcin Wojciech Solarz** is a geographer, political scientist, and the IR researcher at the Faculty of Geography and Regional Studies of the University of Warsaw, Poland; Head of the Department of Political and Historical Geography, FGRS UW; and Head of the project 'Forest Germans (Głuchoniemcy, Walddeutsche): the past and present of forgotten local communities in the Carpathian Foothills,' 2020–2025. His previous works include *The Language of Global Development: A Misleading Geography* (2014), *The Global North-South Atlas: Mapping Global Change* (2020), 'Geography and the world's development divides' in the *Elgar Encyclopedia of Development* (2023), and other works; he is also scientific editor and coauthor of the *Atlas of Poland's Political Geography: Poland in the Modern World* (2018; third place in the 'Atlases' category at the International Cartographic Exhibition in Japan in 2019), *New Geographies of the Globalized World* (2018), and the *Atlas of Poland's Political Geography: Poland in the Modern World: 2022 Perspective* (2022).

# The 1914–2024 War Atlas

## Modernity Deciphered Anew

Marcin Wojciech Solarz

Routledge
Taylor & Francis Group

LONDON AND NEW YORK

Designed cover image: Supreme Court Building, Washington, DC. © Tetra Images, Getty Images ID 97765214.

First published 2025
by Routledge
4 Park Square, Milton Park, Abingdon, Oxon OX14 4RN

and by Routledge
605 Third Avenue, New York, NY 10158

*Routledge is an imprint of the Taylor & Francis Group, an informa business*

*British Library Cataloguing-in-Publication Data*
A catalogue record for this book is available from the British Library

ISBN: 978-1-032-54396-3 (hbk)
ISBN: 978-1-032-52391-0 (pbk)
ISBN: 978-1-003-40644-0 (ebk)

DOI: 10.4324/9781003406440

Typeset in Sabon
by Apex CoVantage, LLC

Once they are over, disastrous events acquire a shape that was not discernible at the time to those whose lives were ruined by them. (. . .) nobody involved in the 1340 Anglo-French naval clash off Sluys knew that the two countries were embarking on a Hundred Years' War.

**Niall Ferguson 2021 (Polish ed. 2022: 92)**

# Contents

# Figures

# Editorial page

Maps and diagrams prepared by:

Izabela Gołębiowska: 8.15, 8.20
Jolanta Korycka-Skorupa: 8.4, 8.17
Tomasz Nowacki: 8.3, 8.10, 8.11, 8.22
Jarosława Talacha: 1.1–1.4, 3.1, 7.1–7.12, 8.6, 8.7, 9.1–9.24, 10.3
Marcin Wereszczyński: 3.2–3.4, 5.1, 5.2, 8.19, 10.7–10.9, 11.1–11.23
Weronika Wnuk: 8.5, 8.8, 8.9, 8.16, 8.18, 8.21
Maciej Zych: 8.1, 8.2, 8.12–8.14, 10.1, 10.2, 10.4–10.6, 10.10–10.30

Translation from Polish and English consultation by Pracownia Językowa Małgorzata Klein.
Translation financed by the Faculty of Geography and Regional Studies, University of Warsaw, Poland. Map preparation financed by the Faculty of Geography and Regional Studies, University of Warsaw, and the IDUB Programme (Excellence Initiative – Research University) of the University of Warsaw, Poland.

# Acknowledgements

This atlas could not have been written and drawn without the assistance, commitment, and goodwill of many people and institutions.

The nucleus and foundation of this book lie in a brief introductory article titled *Political Geography in the Time of a New Hundred Years' War: 1914–2022 and Beyond*, which appeared in a special issue of *Miscellanea Geographica – Regional Studies on Development* (Vol. 26, No. 3, 2022) devoted to political geography. Every fragment of the book that calls this article to mind appears with the consent of the magazine's editorial board.

The credit for the excellent maps goes to the following outstanding cartographers with whom I had the honour to work on this book: Izabela Gołębiowska, Jolanta Korycka-Skorupa, Tomasz Nowacki, Jarosław Talacha, Marcin Wereszczyński, Weronika Wnuk, and Maciej Zych. The atlas would have been noticeably worse without their cartographic skills and commitment.

I also wish to extend my gratitude to Przemysław Śleszyński whose maps, which were unveiled during the official presentation of 'Atlas of Poland's Political Geography. Poland in the Modern World: 2022 Perspective' (which I edited) in the autumn of 2022, inspired the series of maps illustrating the attitude of the international community to the major international issues from the end of WWII to the ongoing Russo-Ukrainian War.

As some of the maps in this atlas were compiled in 2020–2022 for the geographical and political atlas mentioned earlier, I would like to thank the publisher, viz. the Niepodległa Program Office, for their kind permission to use them here.

I would like to thank the Faculty of Geography and Regional Studies and the IDUB (Excellence Initiative – Research University) Programme at the University of Warsaw for financing the drawing up of most of the maps and graphs and the translating the atlas. I would especially like to thank Dr. Sylwia Dudek-Mańkowska, vice-dean for financial and academic affairs at the faculty.

Special thanks must go to the translators and editors at Pracownia Językowa Małgorzata Klein, especially Zuzanna Grzegorczyk and Stephen Canty, who are responsible for the English version of this book.

This atlas certainly contains shortcomings and errors that were not detected during proofreading. I have done what I can to avoid and eliminate them, but in the event that you do discover them, the fault is entirely my own.

# 1     The cogs

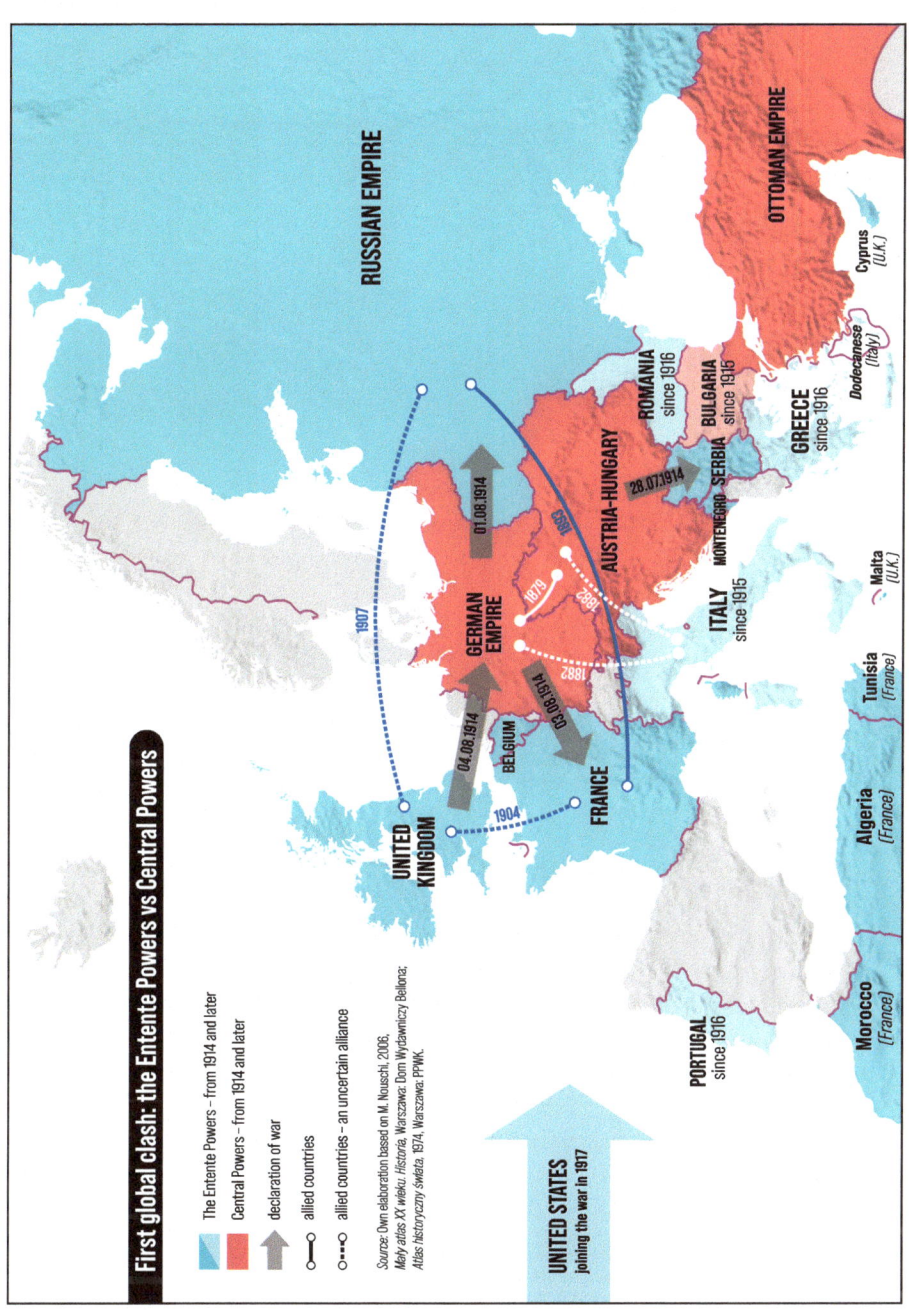

*Figure 1.1* First global clash: the Entente Powers vs the Central Powers

DOI: 10.4324/9781003406440-1

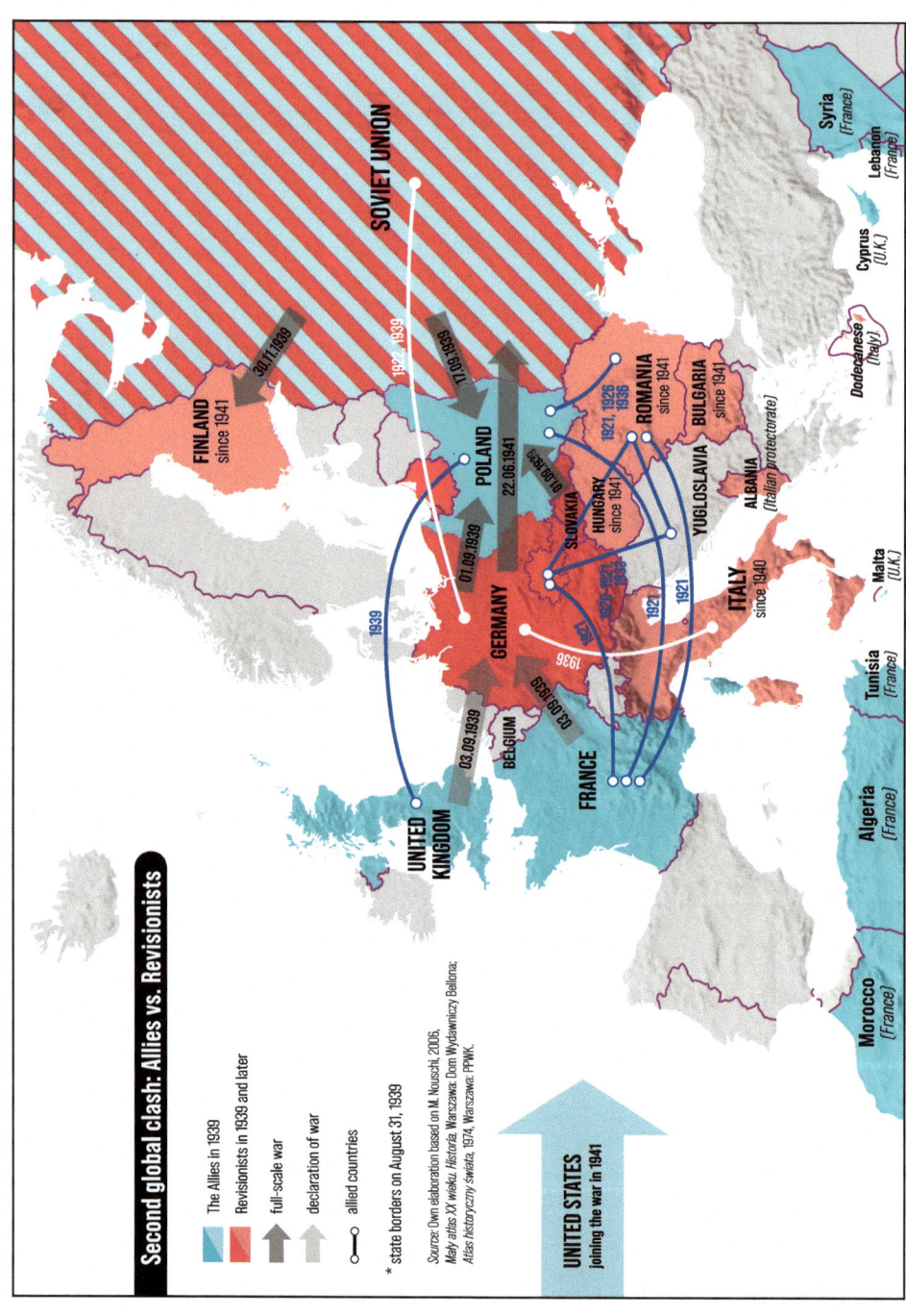

*Figure 1.2* The second global clash: Allies vs Revisionists

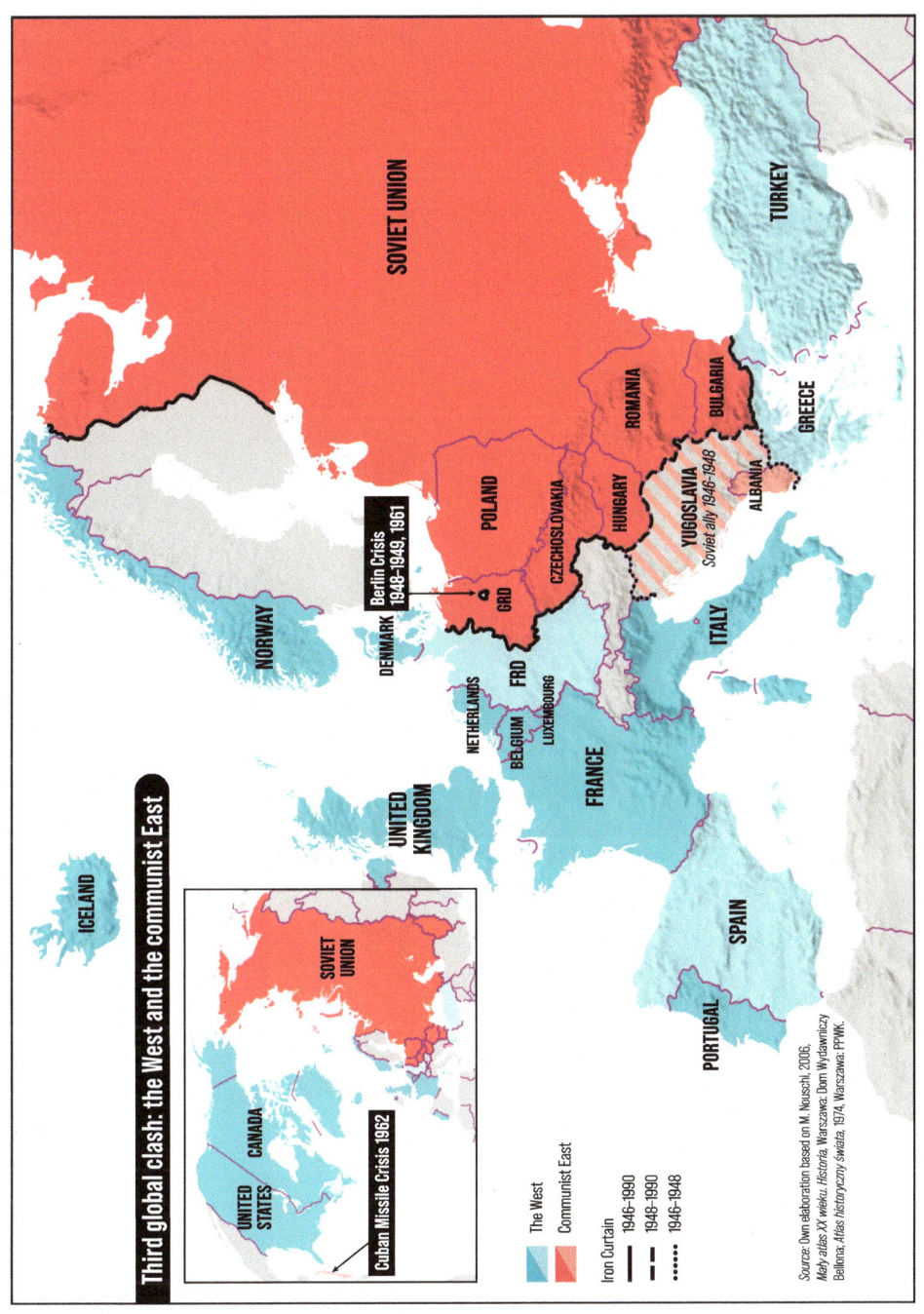

**Third global clash: the West and the communist East**

SOVIET UNION

TURKEY

GREECE

ROMANIA

BULGARIA

ALBANIA

YUGOSLAVIA
*Soviet ally 1946–1948*

POLAND

CZECHOSLOVAKIA

HUNGARY

Berlin Crisis
1948–1949, 1961

GRD

DENMARK

NORWAY

FRD

NETHERLANDS

BELGIUM

LUXEMBOURG

UNITED
KINGDOM

FRANCE

ITALY

SPAIN

PORTUGAL

ICELAND

UNITED
STATES

CANADA

SOVIET
UNION

Cuban Missile Crisis 1962

The West

Communist East

Iron Curtain

1946–1990

1948–1990

1946–1948

*Source: Own elaboration based on M. Nouschi, 2006;*
*Mały atlas XX wieku, Historia, Warszawa: Dom Wydawniczy*
*Bellona; Atlas historyczny świata, 1974, Warszawa: PPWK.*

*Figure 1.3* The third global clash: the West and the communist East

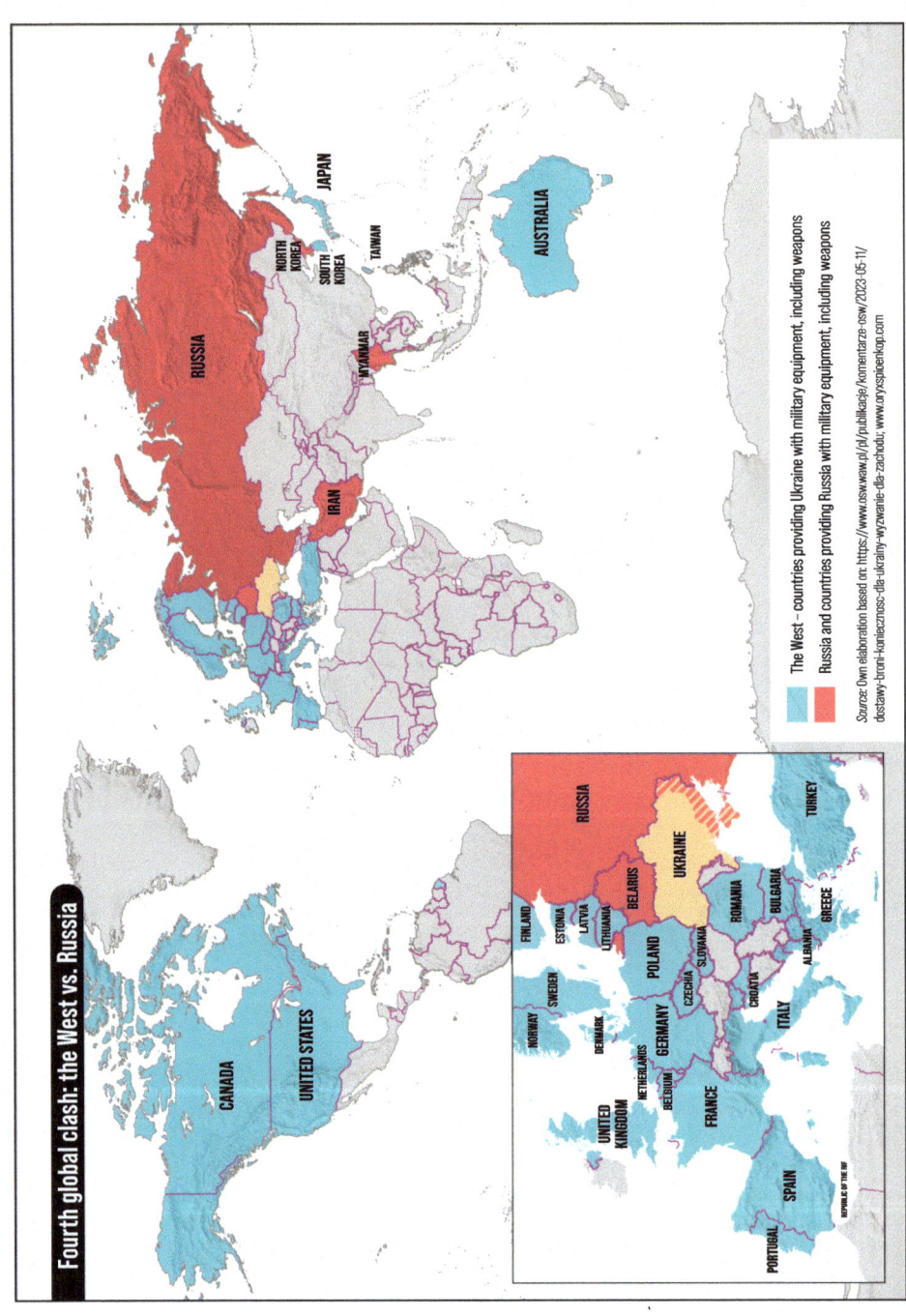

*Figure 1.4* The fourth global clash: the West *vs* Russia

# 2 Reading maps – framing the space

The maps and diagrams in this atlas do more than lend weight to the introductory and explanatory texts by way of illustration. Every one of them is something of a '(carto) graphic essay' in its own right – one that deserves careful reading and reflection and that invites comparisons with other (carto)graphic compilations prepared for this and other books. Maps show connections that may go unnoticed when reading a text (Goldin and Muggah, 2020). Similarly, the maps and diagrams in this atlas can open our eyes to facts, processes, problems, and relationships that might be effectively hidden beneath the sentences of an article or book. In a world where information matters, and even more so, information whose content is 'dense' and whose reception is rapid and in which the image is important, maps continue to convey knowledge and confer power (Blacksell, 2006).

However, when reading maps, and recognising their attractiveness, advantages, and usefulness, it is necessary to be mindful of the hazards involved and remember certain rules. It is easy to lose your way if you are not aware of these risks or fail to observe the rules. The most straightforward rule is to treat a legend as the starting point for reading the map. A map is an attractive information vehicle, but it still has to be read critically. Moreover, it has to be borne in mind that the aims and opinions of the author (and sometimes those of his or her principal) find expression in it (Blacksell, 2006; Goldin and Muggah, 2020). When drawing and reading maps, geography and the associated experience of history are also important because they make us sensitive to some global, regional, and local issues but indifferent to others. In the case of the history of events (to which this atlas is closely related), we are not so much witnesses to it as participants in it, although fate (geography) has assigned us different seats in the theatre of history (unfortunately, the author's country was given a front-row season ticket to every opening night since 1914). Contrary to appearances, a map is not neutral and is not omniscient in its geographical and material scope, as it invariably distorts reality and is always selective in the facts it displays. It always has something that is not seen, but this is not necessarily due to ill will on the part of the cartographer. The map projection, the manner of presentation, the colours used, and the facts selected are always subjective to some extent and are sufficient reason for deficiencies and distortions (Blacksell, 2006; Goldin and Muggah, 2020). Unfortunately, not all the maps in this book contain complete data and comparisons, and some of the maps that were planned could simply not be drawn. After all, maps are merely prisoners of the data on which they are based and without which they cannot be made.

However, the hazards and challenges mentioned earlier do not detract from the attractiveness and usefulness of maps, a language peculiar to geography (they sometimes even

DOI: 10.4324/9781003406440-2

enhance them, especially in politics) (Blacksell, 2006; Goldin and Muggah, 2020). It is my hope that the maps and diagrams in this atlas will make it possible to look at the modern world from a new perspective and that they will encourage reflection and discussion, similarly to introductory and explanatory texts. As a geographer, I stop short of endorsing Michel Houellebecq's assertion that the map is more interesting than the territory (Houellebecq, 2010), but Albert Einstein was absolutely correct when he said: 'You can't use an old map to explore a new world' (Goldin and Muggah, 2020).

# 3   The scenes

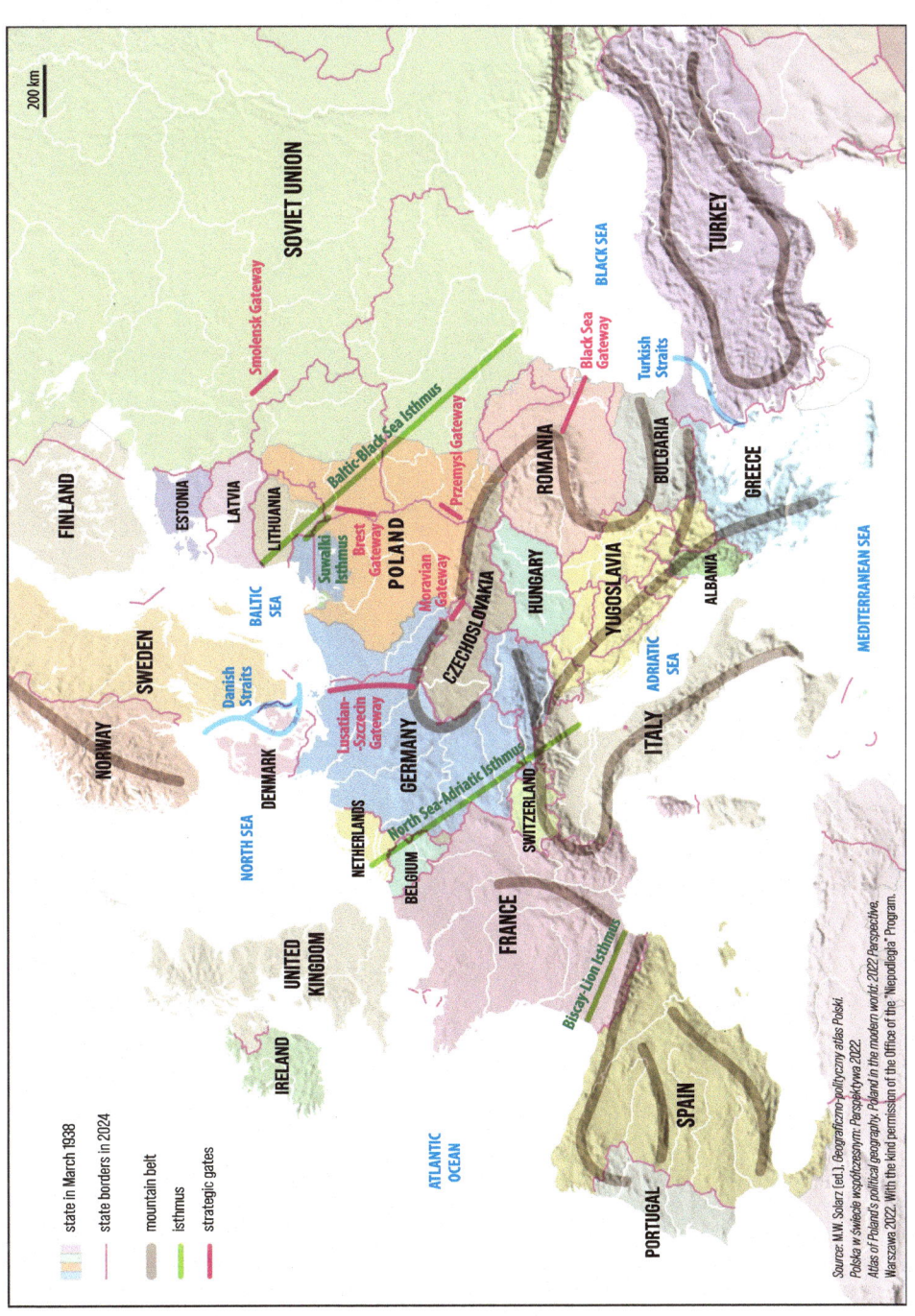

*Figure 3.1* Geographical determinants of European policy

DOI: 10.4324/9781003406440-3

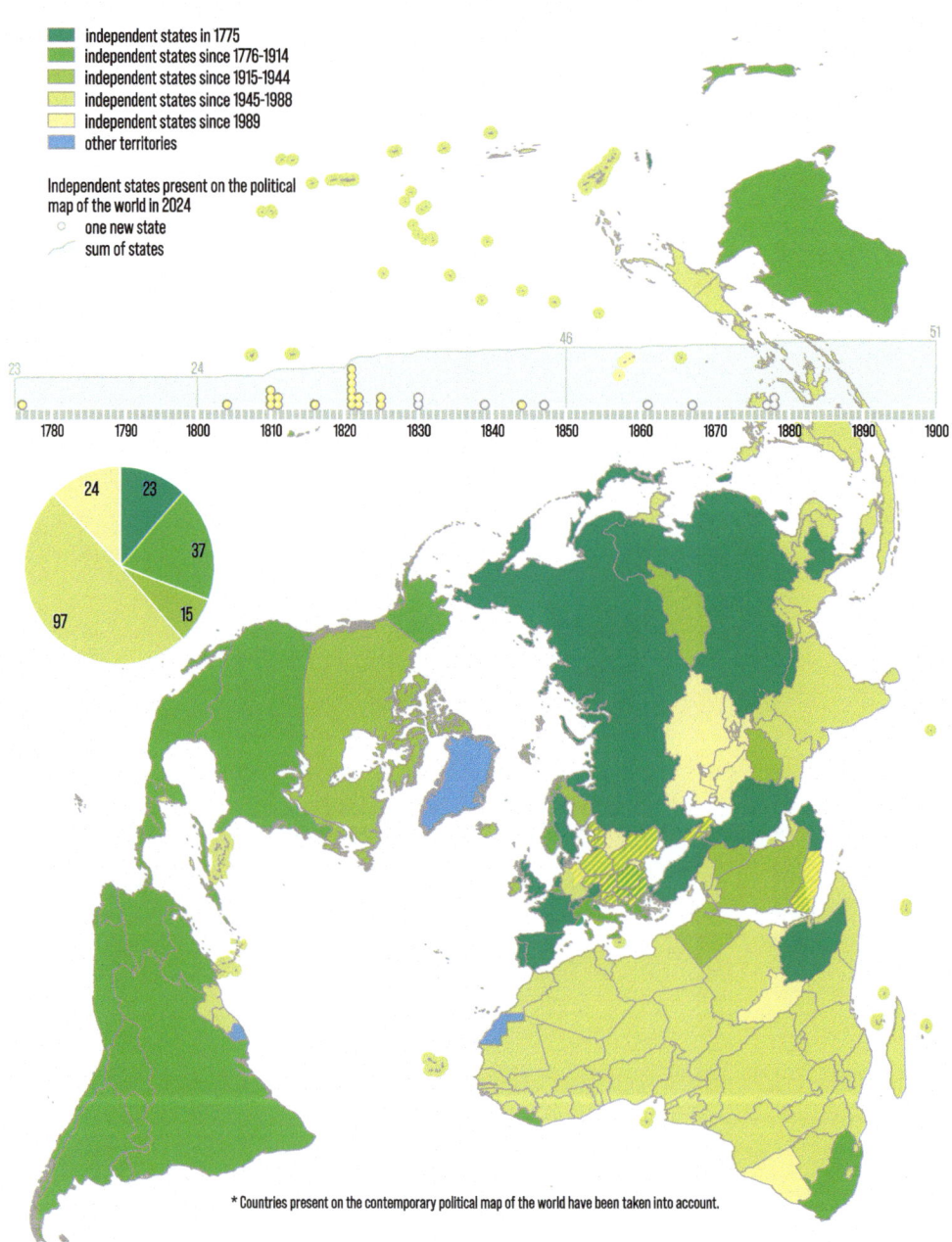

*Figure 3.2* Increase in the number of independent states since 1776

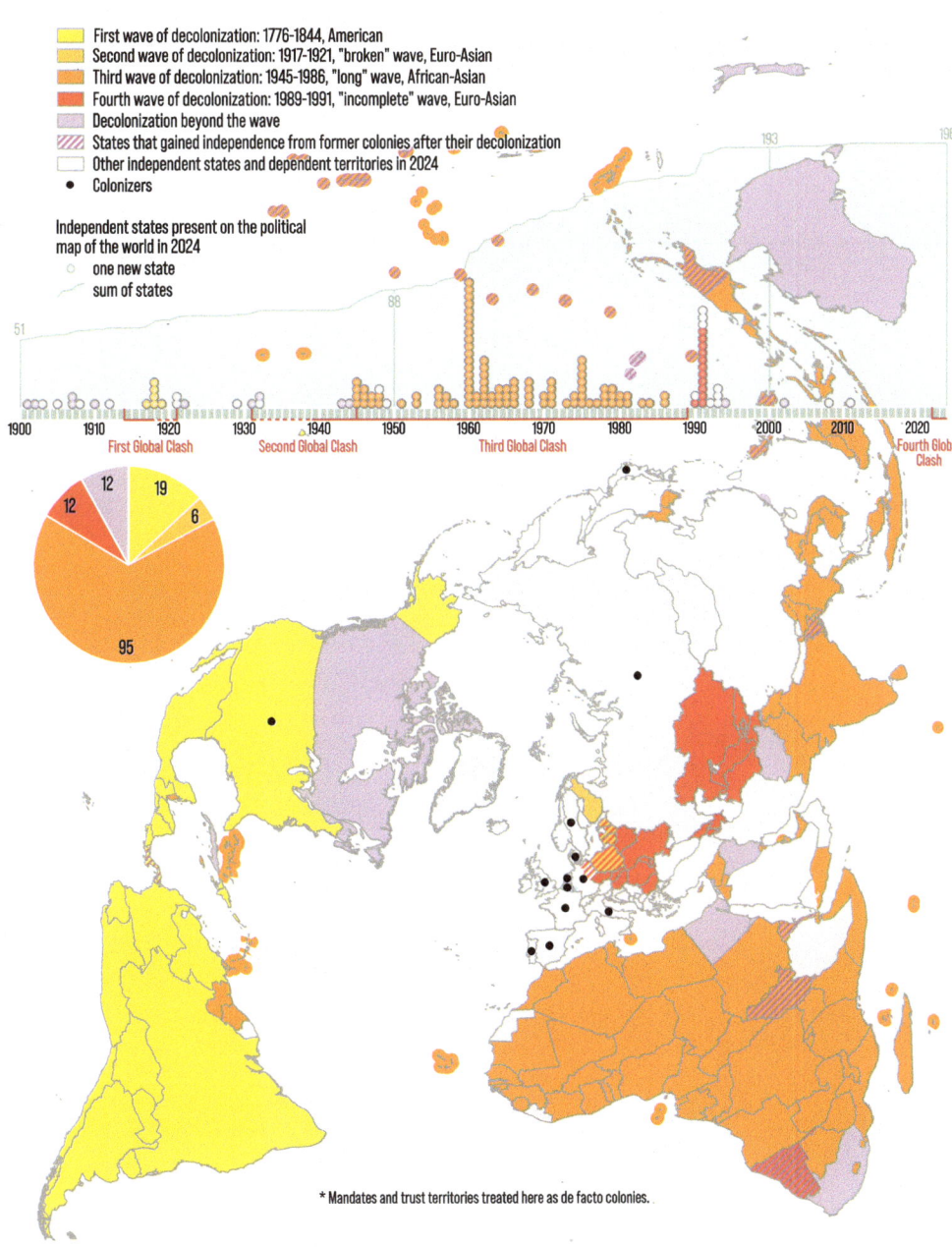

First wave of decolonization: 1776-1844, American
Second wave of decolonization: 1917-1921, "broken" wave, Euro-Asian
Third wave of decolonization: 1945-1986, "long" wave, African-Asian
Fourth wave of decolonization: 1989-1991, "incomplete" wave, Euro-Asian
Decolonization beyond the wave
States that gained independence from former colonies after their decolonization
Other independent states and dependent territories in 2024
● Colonizers

Independent states present on the political
map of the world in 2024
○ one new state
sum of states

First Global Clash   Second Global Clash   Third Global Clash   Fourth Global Clash

* Mandates and trust territories treated here as de facto colonies.

*Source:* Own elaboration based on: The World Factbook, https://www.cia.gov/the-world-factbook/

*Figure 3.3* Waves of decolonisation: 1776–2024

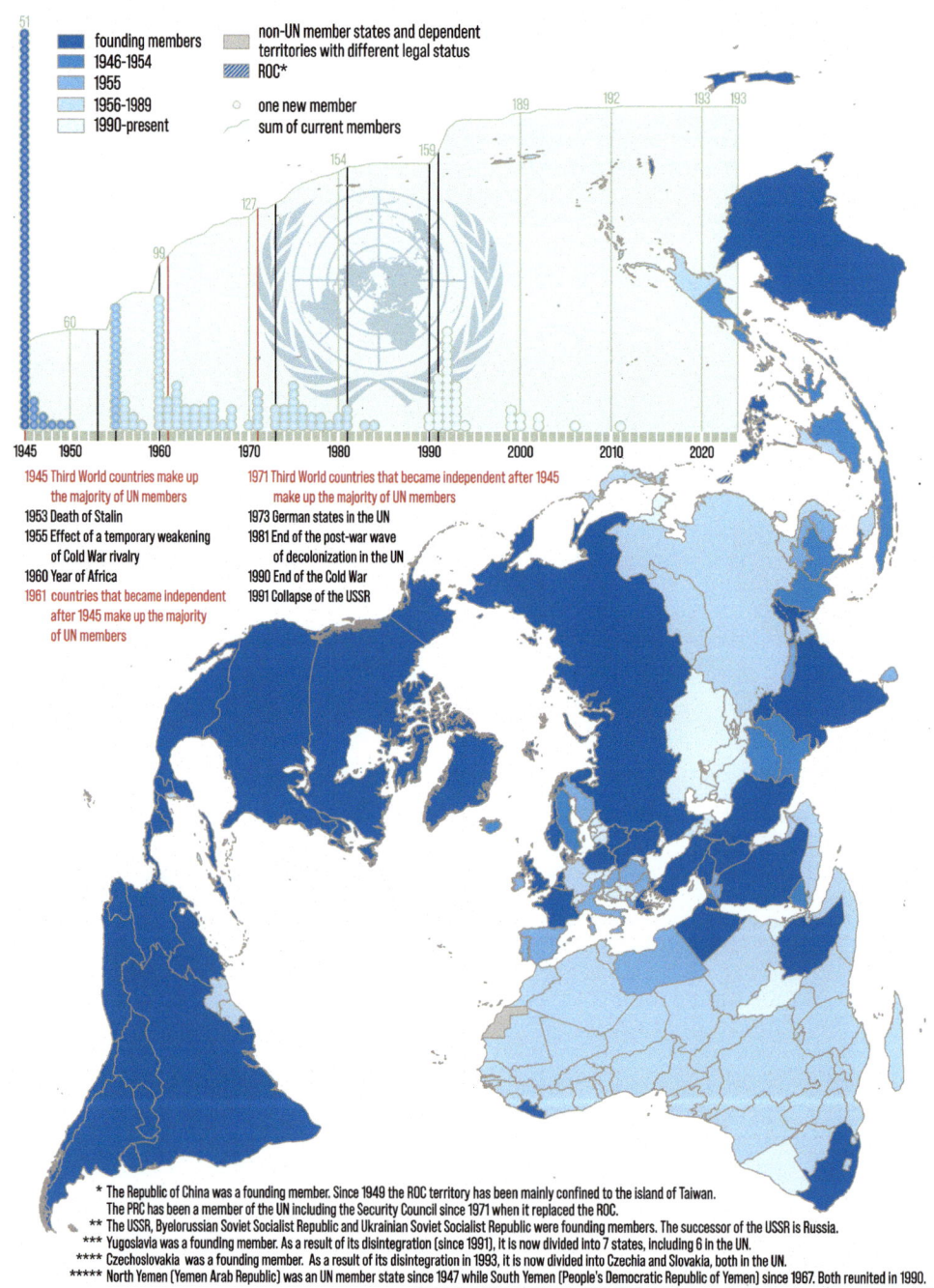

*Figure 3.4* United Nations membership: 1945–2024

# 4 Deciphering history – framing the time

In 1974, Juliusz Mieroszewski, a political émigré from communist Poland, observed that

> Certain situational arrangements and patterns have recurred throughout history. In most cases, however, history is no more than a catalogue of opening nights. History is fascinating, because 'the same thing' is never 'the same thing' and a new occurrence of a practically identically situation can produce different results.
>
> (Mieroszewski, 1974, p. 3)

By their very nature, WWI, WWII, the Cold War, and the current Russo-Ukrainian War are calamitous 'opening nights' performed on a global stage. At the same time, however, they form a certain 'situational arrangement and pattern' that spanned the hundred years from 1914. The 1914 July Crisis set in motion a mechanism whose first cog drove the next, which became the impulse that set in motion a sequence of others that will continue until the last of them has occurred.

History is made up of stories with different rhythms. It operates in varying chronological realities, moving between the instant of time and a long-term history that spans centuries (Braudel, 1971). People feel closest to a fast-paced story about the world around them; one told on the run, with shallow breath – in short, a history of events, but one that tends to fade rapidly in the sunlight of history (Braudel, 1971). Professional storytellers and researchers of the history of events, as well as the mass media and social media, are immersed in the fast-moving current of history (Kapuściński, 2007), attempting to explain it while hemmed in by their own entanglement in history and geography. It is not surprising then that many of the narratives about modernity quickly become obsolete and make way for new ones. Alongside the history of events, however, there is a more profound history of conjuncture that utilises broad timespans covering bunches of decades (Braudel, 1971). This enables history to be described from a further perspective, viz. from the banks of the river of history, in a bid to additionally uncover its 'situational arrangements and patterns' – to the extent that they exist. A story of conjuncture, i.e., one that spans several decades and conjoins many critical events, may be a closed story of a cyclist covering a new route at each stage of a road race or of a cyclist completing successive laps of a velodrome, where the same short route is repeatedly covered but under different circumstances. In this way, the history of conjuncture becomes a history of relatively short cycles revealing 'situational arrangements and patterns'. Braudel's vision of history, however, incorporates an even more profound narrative – a history covering a time span measured in centuries (Braudel, 1971). In contradistinction to the history of events and of conjuncture (including short cycles), Braudel's history of the *longue durée* is not narrated

DOI: 10.4324/9781003406440-4

from the current of history or from its sidelines, but from a high hill towering over the course of history and from where the entirety of the historical process can be taken in. The latter perspective, however, is too long for the story told in this atlas. The view proposed here is a combination of the history of events and the history of short cycles. The atlas looks at the two world wars and the Cold War from the sidelines of history, from closer up or further back, but the Russo-Ukrainian War completely lacks any time perspective. At the time of writing, this is still a story told with shallow breath, without the distance of time, and with incomplete knowledge of its causes, course, results, and effects.

Eric Hobsbawm (1994) proceeded similarly some 30 years ago. He discerned that the years 1914–1991 stood out from the rest of the 20th century and called this period a 'Short 20th Century', even though this was still part of the history of events for the British historian (Hobsbawm, 1994). As he published his synthesis of 20th-century history in 1994, he was looking at the end of the Cold War and the fall of the Soviet Union from the standpoint of a participant in, as well as a witness of, history in the making. The conscious error (there being no alternative) of his overly short perspective, which resulted from his vantage point on the timeline, led to him to assert without serious doubt that the turn of the 1990s ended a coherent era in world history and began a new one (Hobsbawm, 1994). However, the 30 years that have elapsed since these words were written and the Russo-Ukrainian War (since 2022) call for a reappraisal of the years 1989–1991. It now appears that they did not bring the historical process that began in 1914 to an end. Revising Hobsbawm's view reduces the significance of the turn of the 1990s. This, however, again reflects the perspective of a history of events, which would risk committing every conceivable error made possible by the lack of an appropriate historical (and in this case also geographical) perspective, i.e., a history that could not possibly take into account crucial but as yet unknown facts, processes, or circumstances and that might, in the case of known ones, place excessive or insufficient weight on them or misplace them in the overall context. Hobsbawm goes on to point out that despite his or her involvement in the history of events, the scholar can nevertheless make some sense of the history in which he or she is a participating observer (and this has to be hoped for when studying without a historical perspective), provided that he or she first and foremost observes it accurately and listens to its reverberations (Hobsbawm, 1994). The history in this atlas is similarly narrated and interpreted from the standpoint of the participating observer, who accurately observes history and listens to its reverberations as much as possible. As for the places where many of the important events of world history between 1914 and 2024 have taken place, it is seen from a perspective of the Central European viewer who has a front-row seat in the theatre of modern history.

# 5 The milestones

*Figure 5.1* Milestones of the new Hundred Years' War vs GDP

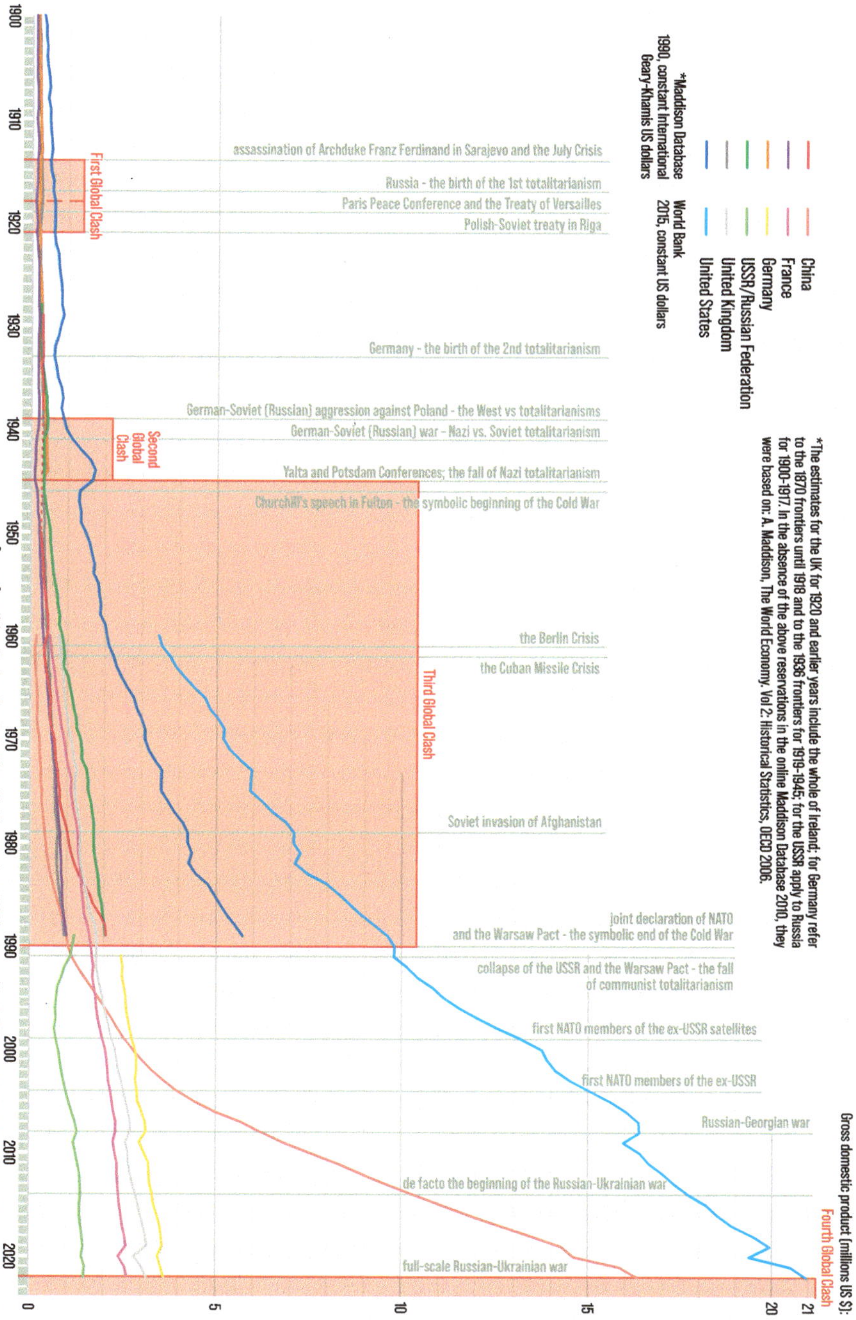

*Source:* Own elaboration based on: A. Maddison, Maddison Database 2010 (www.rug.nl/ggdc/). The World Bank data (https://databank.worldbank.org)

DOI: 10.4324/9781003406440-5

*Figure 5.2* Milestones of the new Hundred Years' War vs per capita GDP

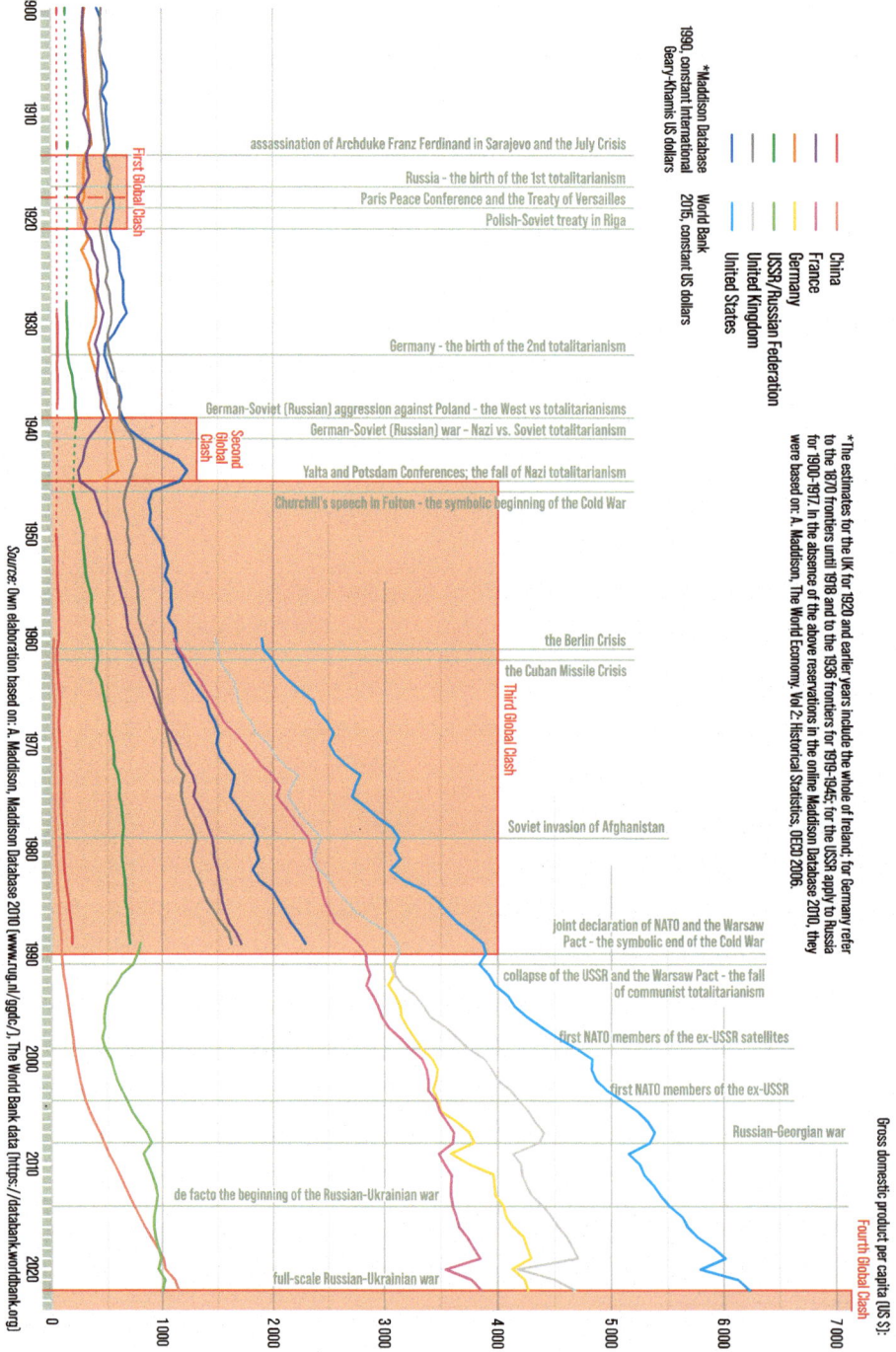

# 6 The first cog – European cycle of power and decay vs the new Hundred Years' War

Paul Kennedy considers that, until around 1500, the development levels of the great centres of civilisation, viz. the Ottoman Empire, Ming Dynasty China, Mughal India, and Europe, can accurately be described as roughly equivalent. He goes on to add that it was by no means obvious at the time that the Europeans would achieve world domination over the course of the next 400-odd years (Kennedy, 1995, pp. 19–20; see also other authors: Milewski, 2004, p. 149). However, in the *Nicomachean Ethics*, Aristotle astutely observes that it was not the most attractive and the strongest who were crowned in the Olympic Games, but those who competed (Aristotle, 2004). The Europeans might not have been the most attractive or the strongest of the world's great civilisations at the time, but they alone took development to the next level while the Chinese, Indian, and Islamic civilisations stagnated (Kennedy, 1995, pp. 20–28).

The European domination of world politics symbolically begins as early as 1756–1763, i.e., with the onset of the Industrial Revolution (symbolically dated from 1763, when James Watt improved on Thomas Newcomen's steam engine) and the outbreak of the first global conflict, viz. the Seven Years' War (1756–1763) (Baugh, 2021, p. 19). However, it is the culmination of processes that began in Europe around 1000 AD when, after five centuries of stagnation, it embarked on a path of accelerated development (Kennedy, 1995, pp. 31–44; Milewski, 2004, pp. 133–186). This introductory phase of the European march towards domination, which lasted until approximately 1756–1763, was characterised by an increase in capabilities and opportunities, by moving to the forefront of the world's civilisations, and, eventually, by revealing European capabilities and ambitions (witness the great geographic discoveries and colonialism). The cumulative changes in the mid-18th century reached a tipping point that constituted a milestone in world history. Along with the Industrial Revolution, as David Landes briefly but emphatically remarked, 'the world had slipped its moorings' (Landes, 1998, p. 192), or to put it more accurately, Europe and its main settler colonies, especially those in North America, which gave them temporary power over the world. Europeans occupied or controlled 35% of the world's land area in 1800, 67% in 1878, and over 84% in 1914 (Kennedy, 1995, p. 155). In 1900, the five major belligerents of WWI, viz. the UK, France, Germany, Russia, and the United States, controlled approximately 44% of the world's land area (Barbag, 1978, pp. 222–223). When reflecting on world history since c. 1500 AD, Zygmunt Bauman's observation that 'as far as history was concerned, Europe was definitely an exporting country' is strikingly apt (Bauman, 2004, p. 8). This unipolar world order, centred around Europe, continued uninterrupted until 1914. This year marked the beginning of the end of the Eurocentric world.

DOI: 10.4324/9781003406440-6

Europe had reached the zenith of its power by mid-1914, but was then devastated by WWI (1914–1918) and WWII (1939–1945). Two rival political and military blocs faced off in 1914, viz. the Central Powers (led by Germany and Austria-Hungary) and the Allied Powers (under the hegemony of France, the UK, and Russia). Germany was removed from the front rank of world powers in 1945 after its defeat in both world wars, while Austria-Hungary failed to survive the first and was broken up into several successor states in 1918. Defeated Russia withdrew from WWI in 1918 and was again defeated in the Polish-Soviet War (1919–1921), which was fought during the ensuing Russian Civil War (1917–1922). The country was plunged into the devastating and murderous maelstrom of a communist revolution in 1917 that resulted in the creation of a bogus global centre of power and development which lasted close on 75 years. Communism had global designs from the outset, but until the mid-1950s, when Khrushchev opened up the Soviet Union to the Global South, the spread of communism was mostly limited to the 'front-line' states of Europe and Asia (Kennedy, 1995; Kanet, 1989; Fukuyama, 1987; Solarz, 2014b). For their part, the UK and France were among the victors of both world wars, but both have experienced a steady erosion of power since 1914. These two European powers could still decide the fate of the world between them in 1919, but by 1945, France no longer had any influence on the course of major world affairs, and the UK was the weakest of the Big Three. The centre of power moved from Europe to the United States in 1914–1956 as a result of both the development of the latter and the destruction wreaked by two world wars on the former.

Obviously, Europe did not step down overnight. The metaphorical curtain finally fell on European domination of international relations with the forced and humiliating withdrawal of British and French forces (at the behest of the United States and the USSR) from the Suez Canal in 1956, but this was merely the last of a sequence of symbolic milestones that showed that Europe was systematically and stubbornly abdicating the role it had played in international relations. In 1931, the UK devalued the pound, thereby bringing about the collapse of the currency system (Zabielski, 1994, p. 39). France began by abandoning Czechoslovakia and Poland – its main Central and Eastern European (CEE) allies in the event of an armed conflict with Germany – in 1938–1939 and then suffered a humiliating military defeat in 1940. Significantly, France was unable to impose the sort of coalition in 1918–1939 that was needed to counter the serious and growing threat from Germany, and which would have benefitted all parties, on Prague and Warsaw. France's disastrous performance in WWII resulted in the country being shut out of the new power circle, viz. the Big Three. In the second half of the 1940s, while the Greek Civil War was raging and the Soviet Union was pressuring Turkey over the Turkish Straits and demanding territorial concessions in the border region with the USSR, the United States stepped into the political and security breach left by the UK (Kennedy, 1995, pp. 368, 372). Soon afterwards, a succession of events in the Global South, then within the sphere of influence of either the UK or France, unequivocally served notice that these two powers were on the way out. The rising tide of decolonisation after 1945 was yet another clear signal of changes in the world order. In just 70 years from 1914, a 'tidal wave of color' (Kelley, 1999; Solarz, 2014b, p. 83) swamped the monochromatic political map of a world that had been dominated by Europe for centuries.

# 7 The European tides

Figure 7.1 Borders in CE Europe: 1815

DOI: 10.4324/9781003406440-7

*Figure 7.2* Borders in CE Europe: June 1914

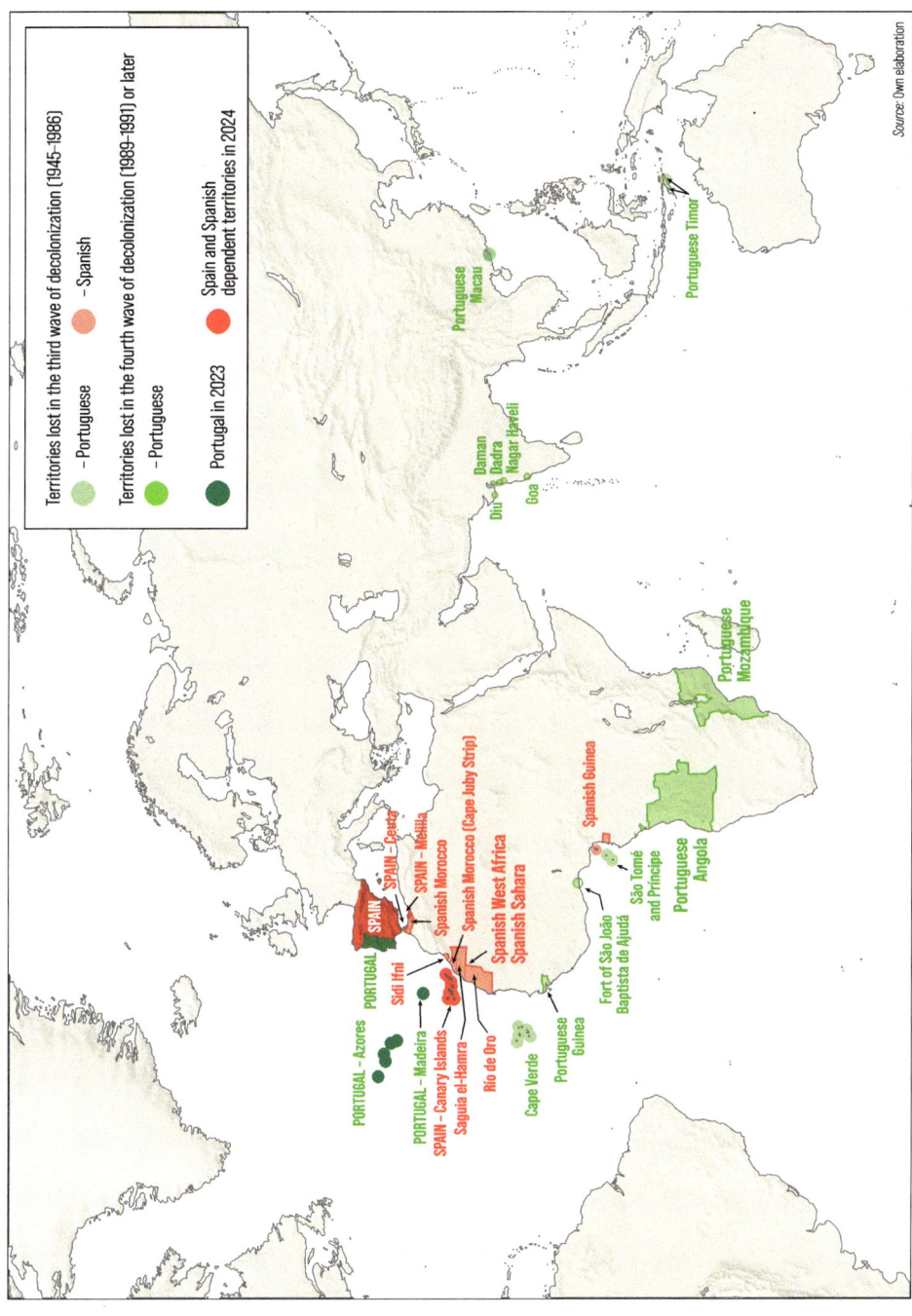

*Figure 7.3* Spanish and Portuguese colonies

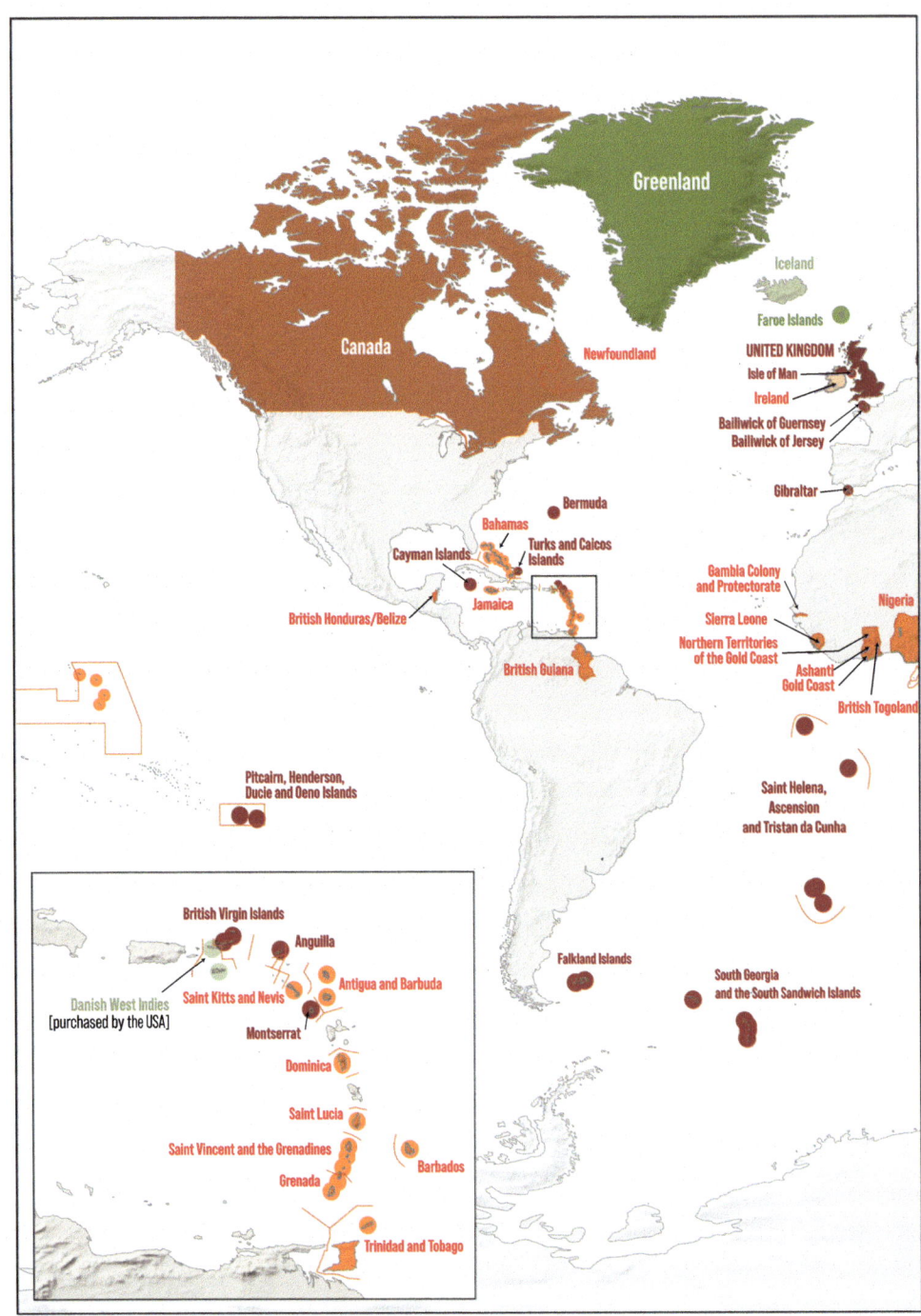

*Figure 7.4* British and Danish colonies

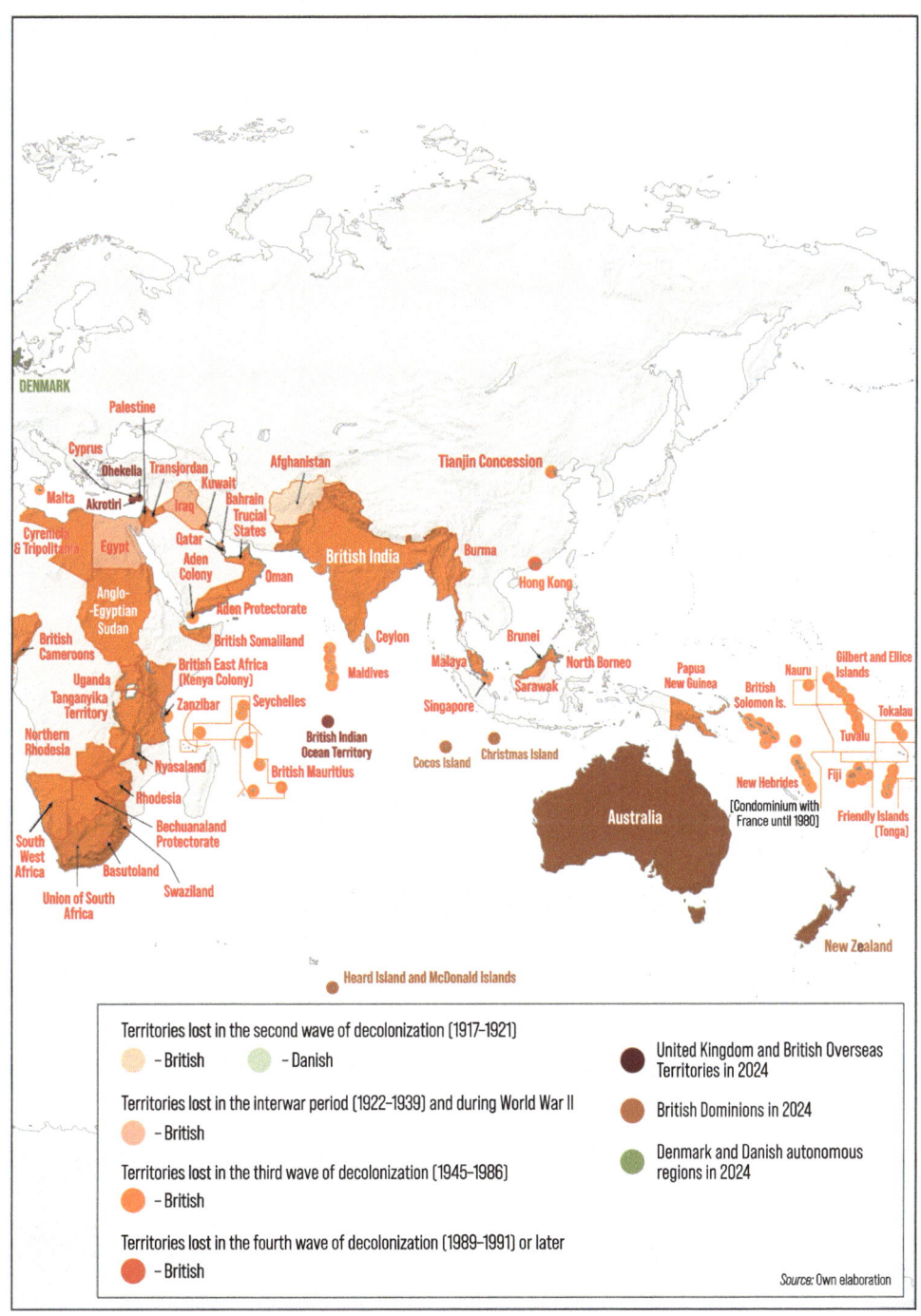

*Figure 7.4* (Continued)

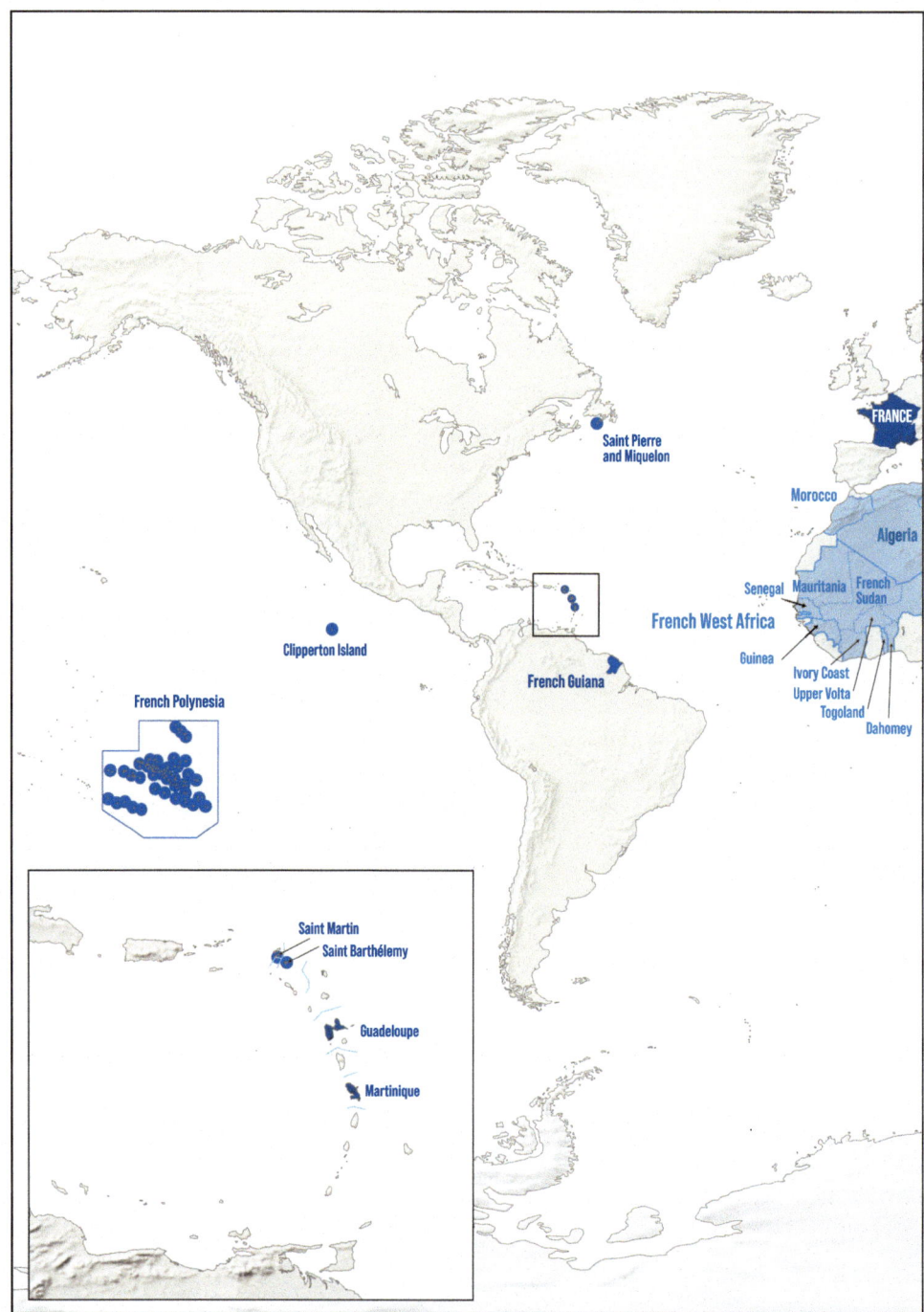

*Figure 7.5* French colonies

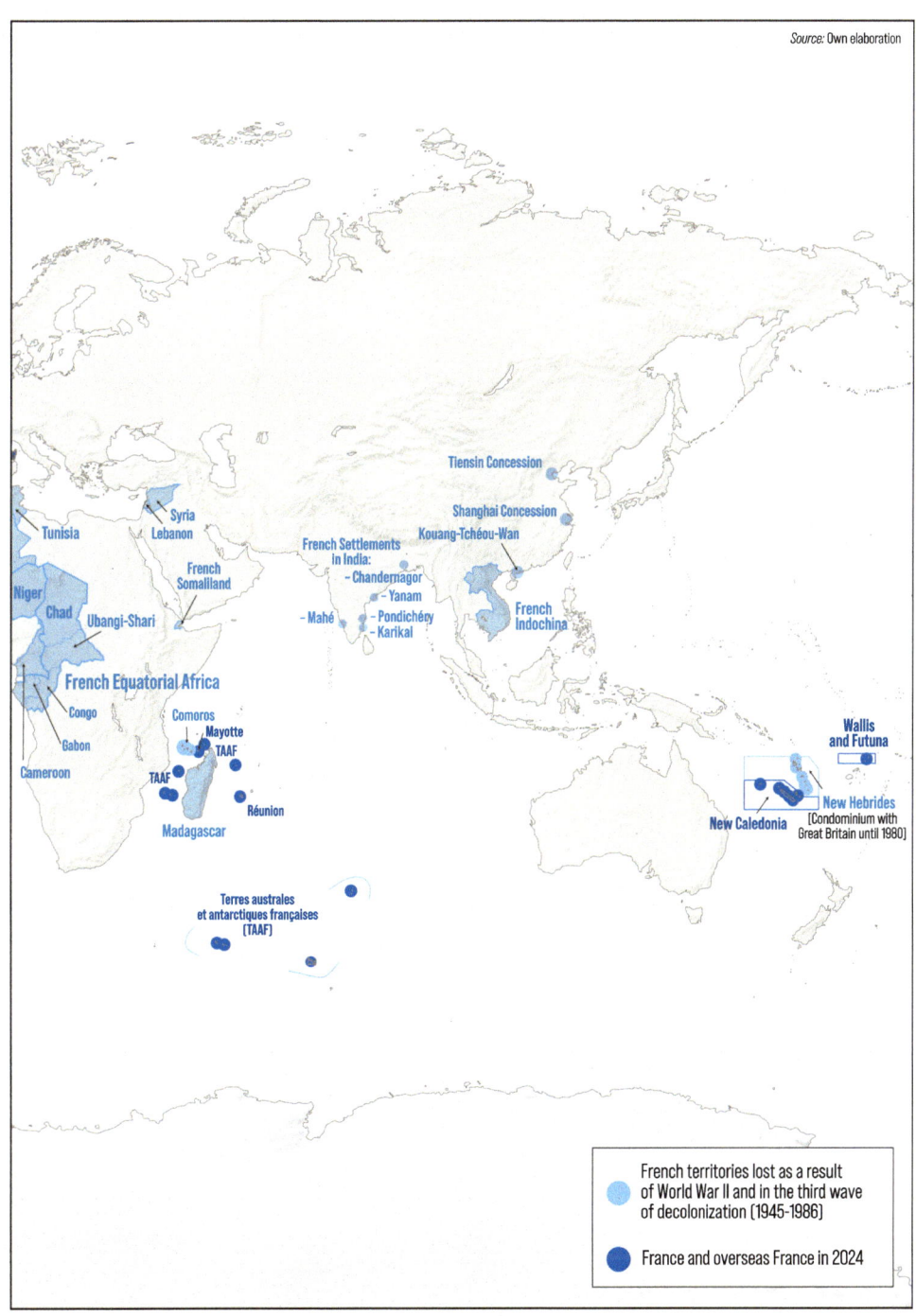

*Figure 7.5* (Continued)

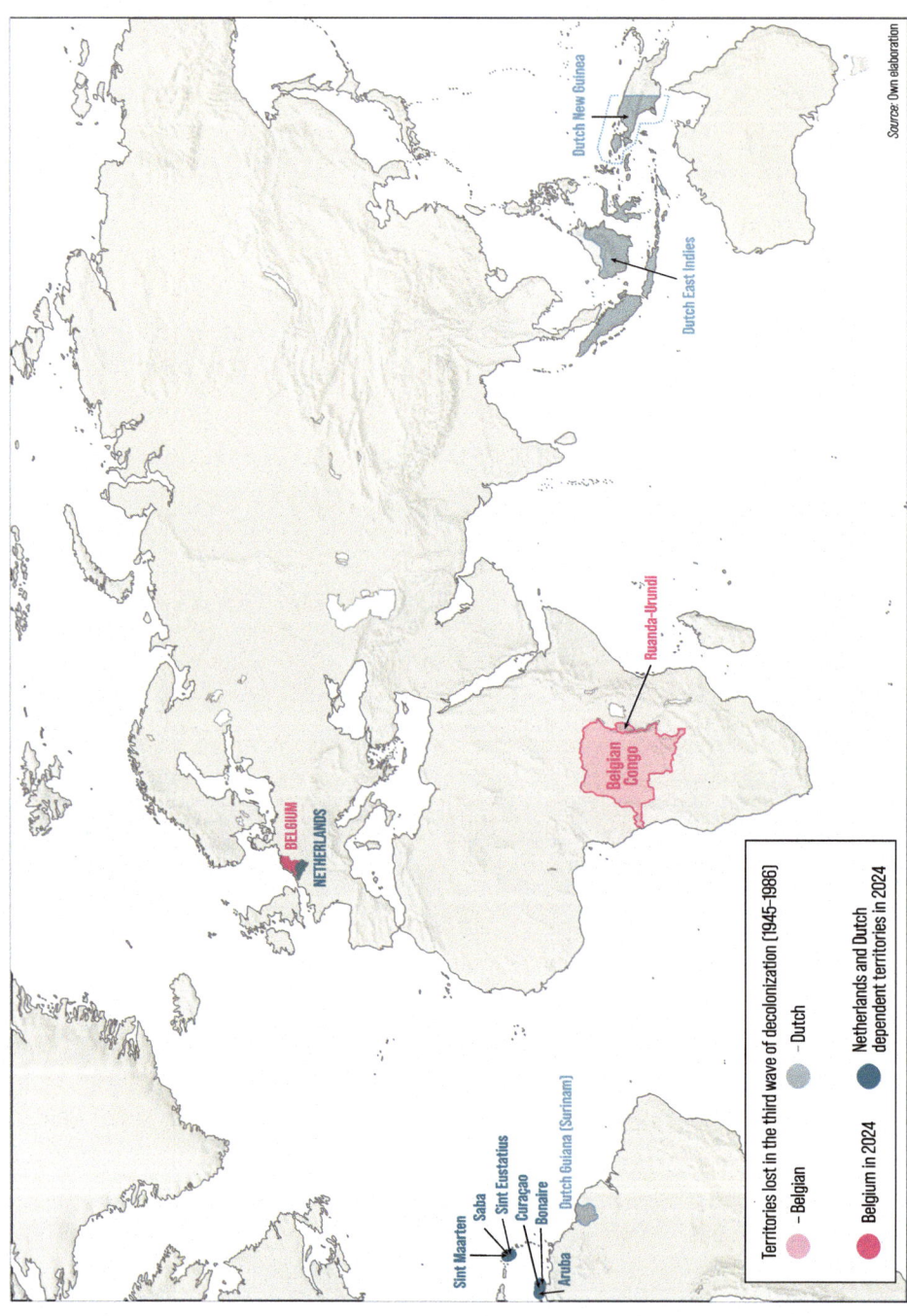

*Figure 7.6*  Dutch and Belgian colonies

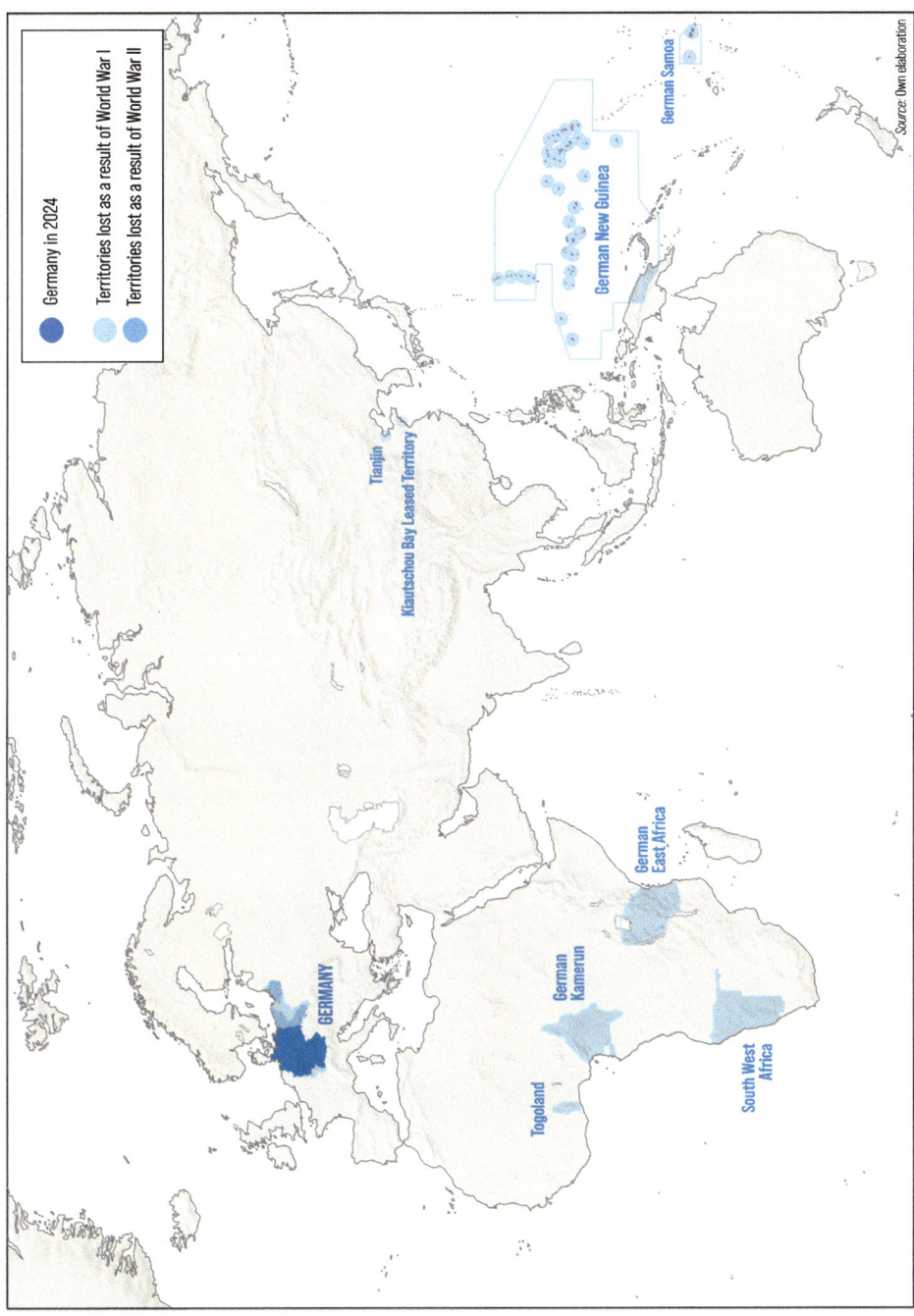

*Figure 7.7* German colonies

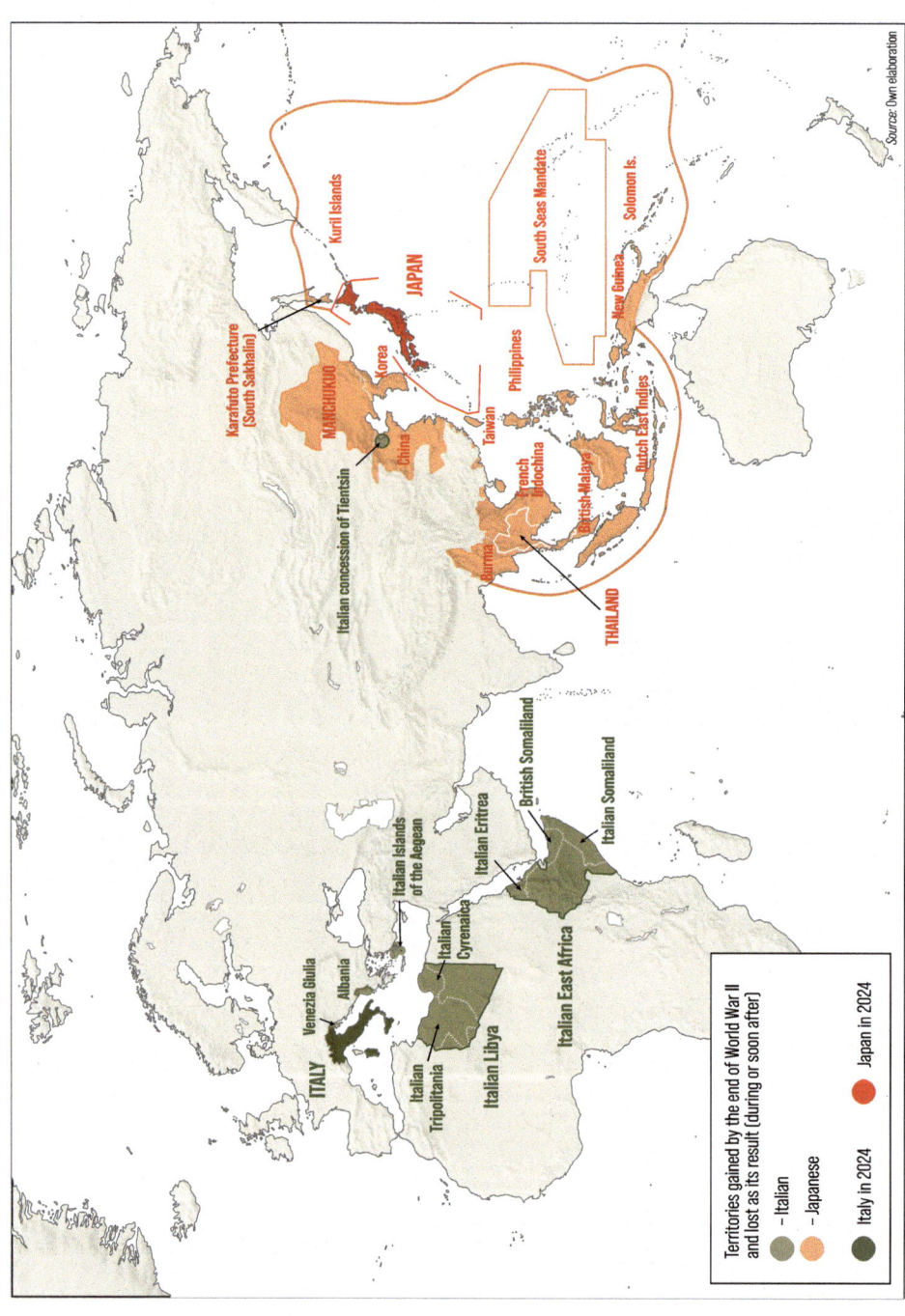

*Figure 7.8* Italian and Japanese colonies

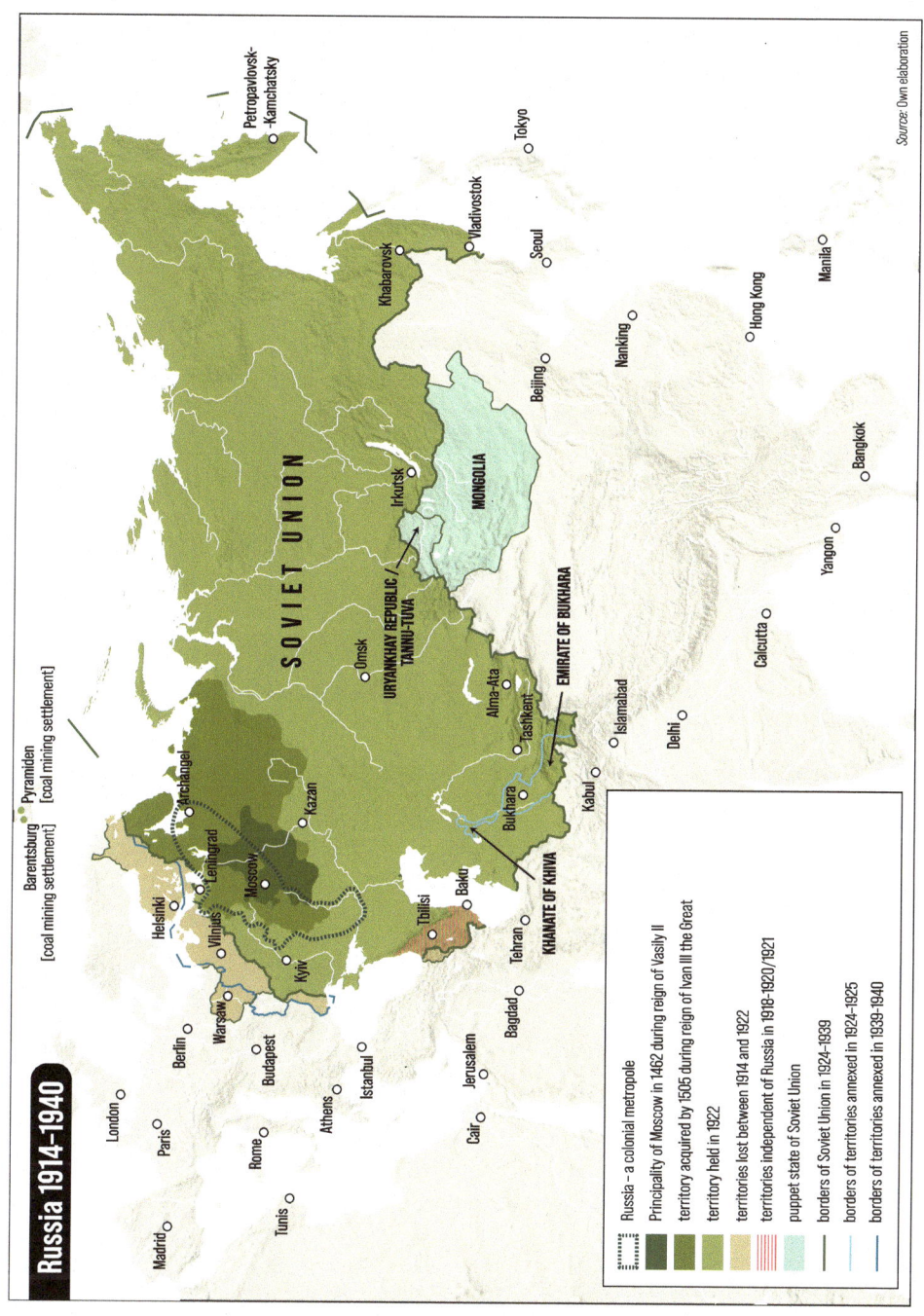

*Figure 7.9* White and Red Russia vs colonies: 1914–1940

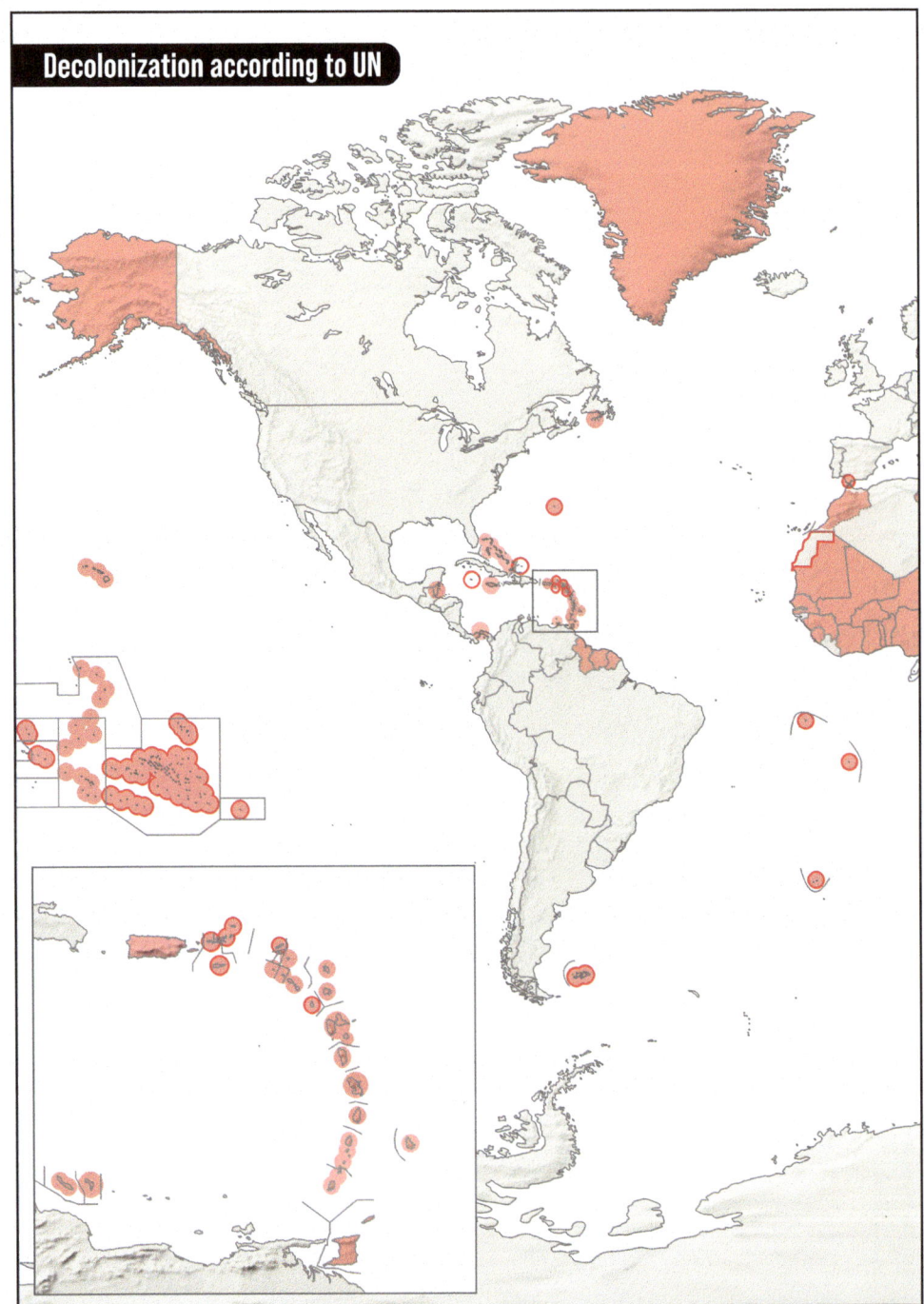

*Figure 7.10* Dependent territories to decolonise, according to the UN, in 1946 and 2024

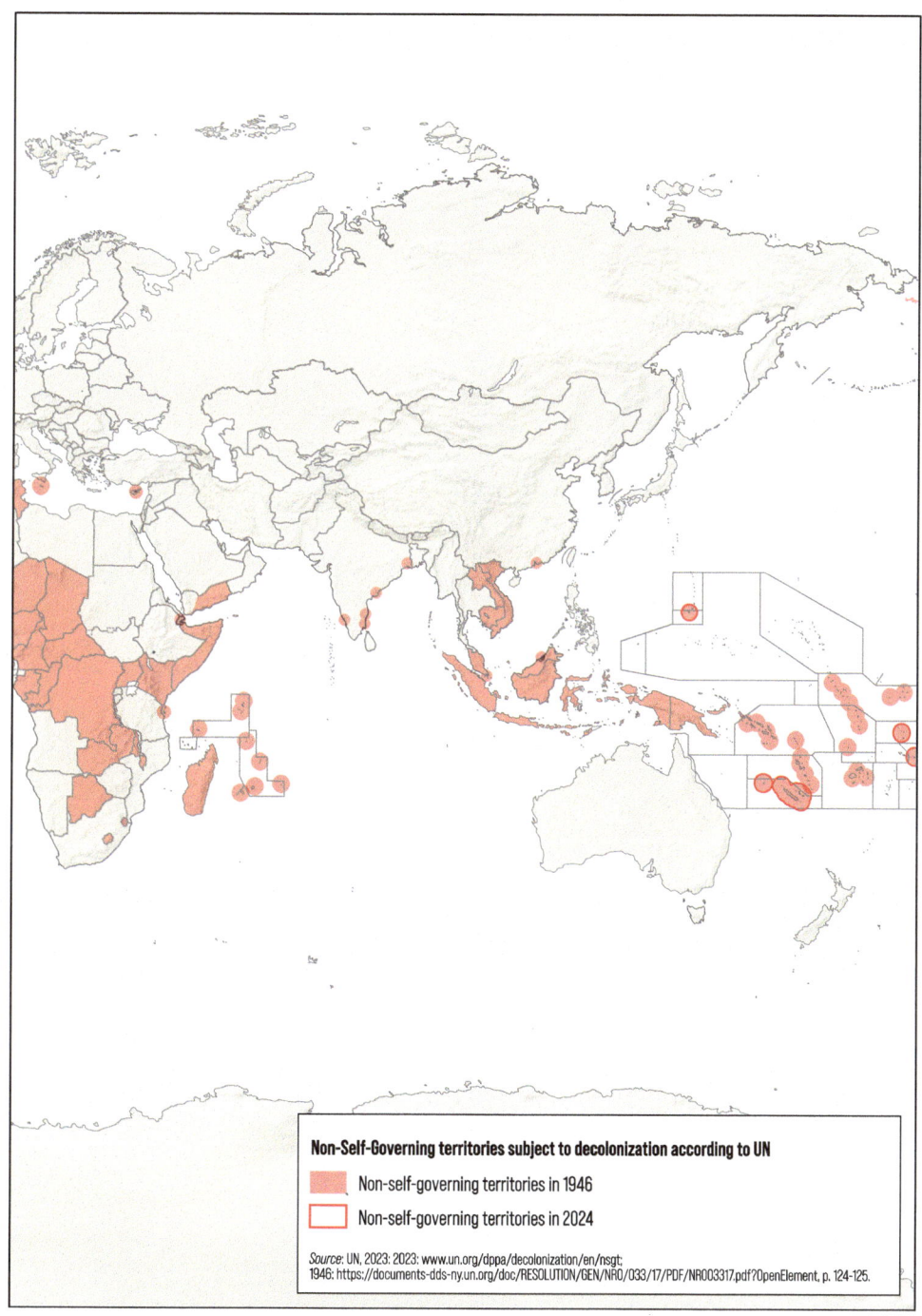

*Figure 7.10* (Continued)

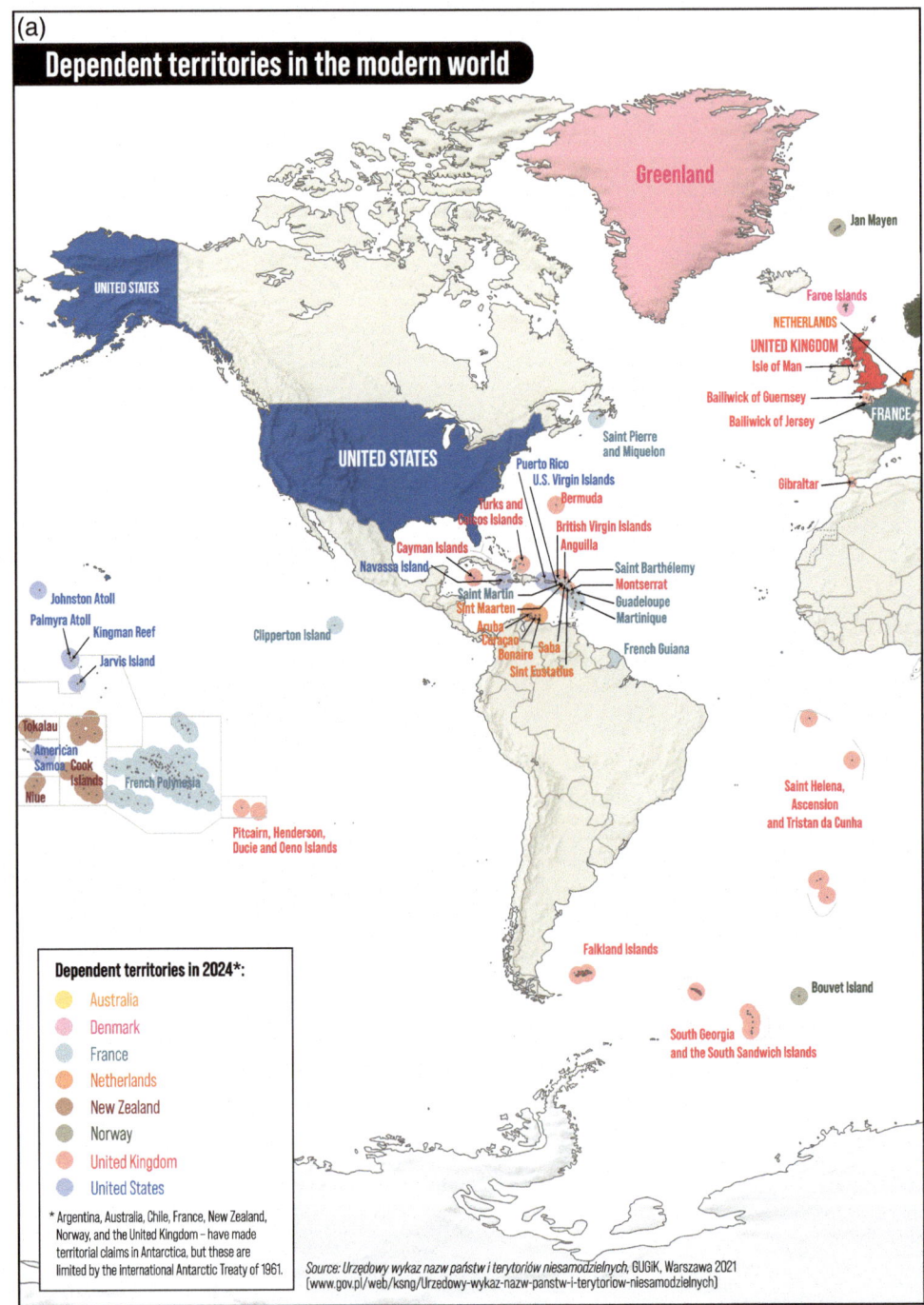

*Figure 7.11* Contemporary dependencies

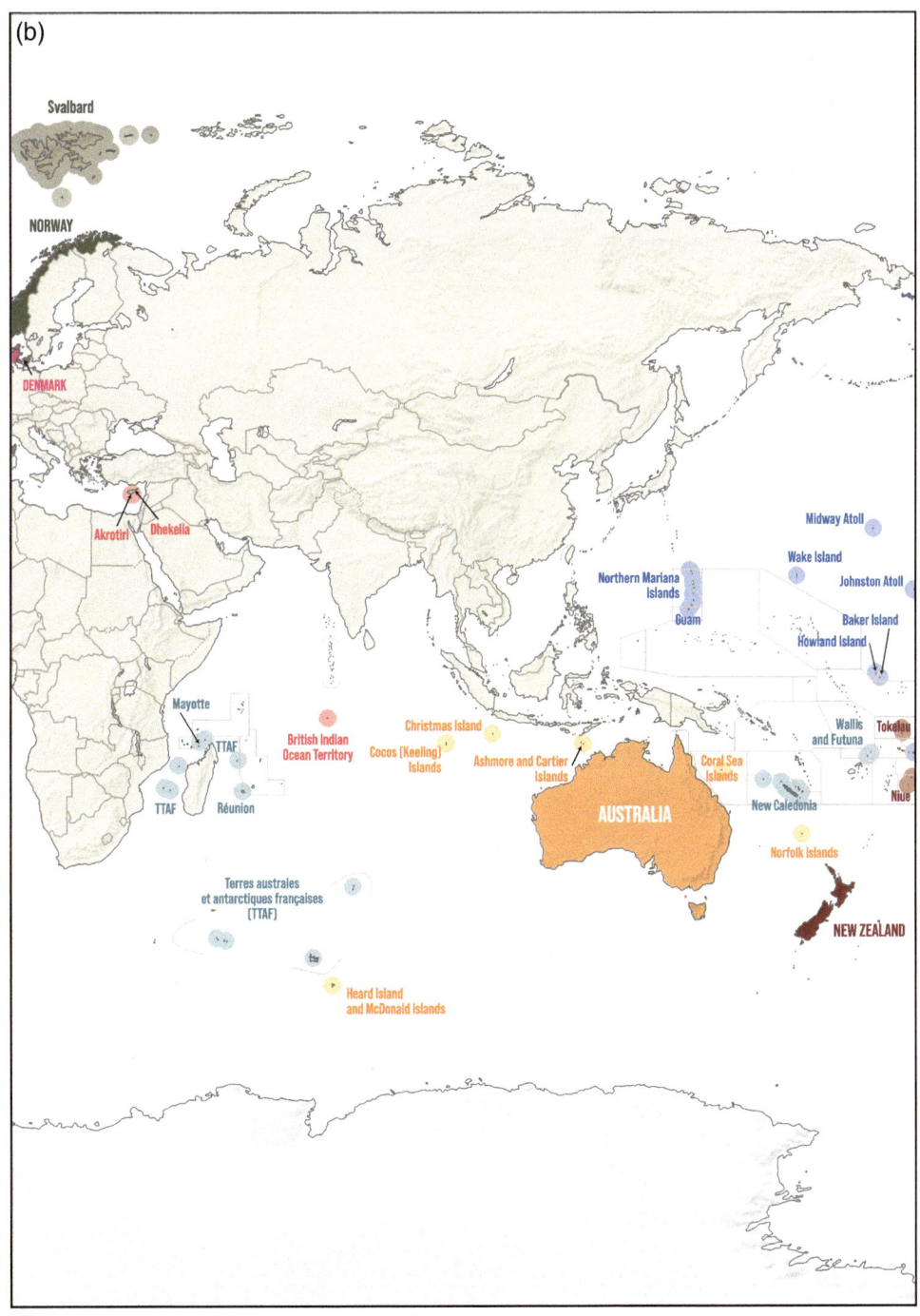

*Figure 7.11* (Continued)

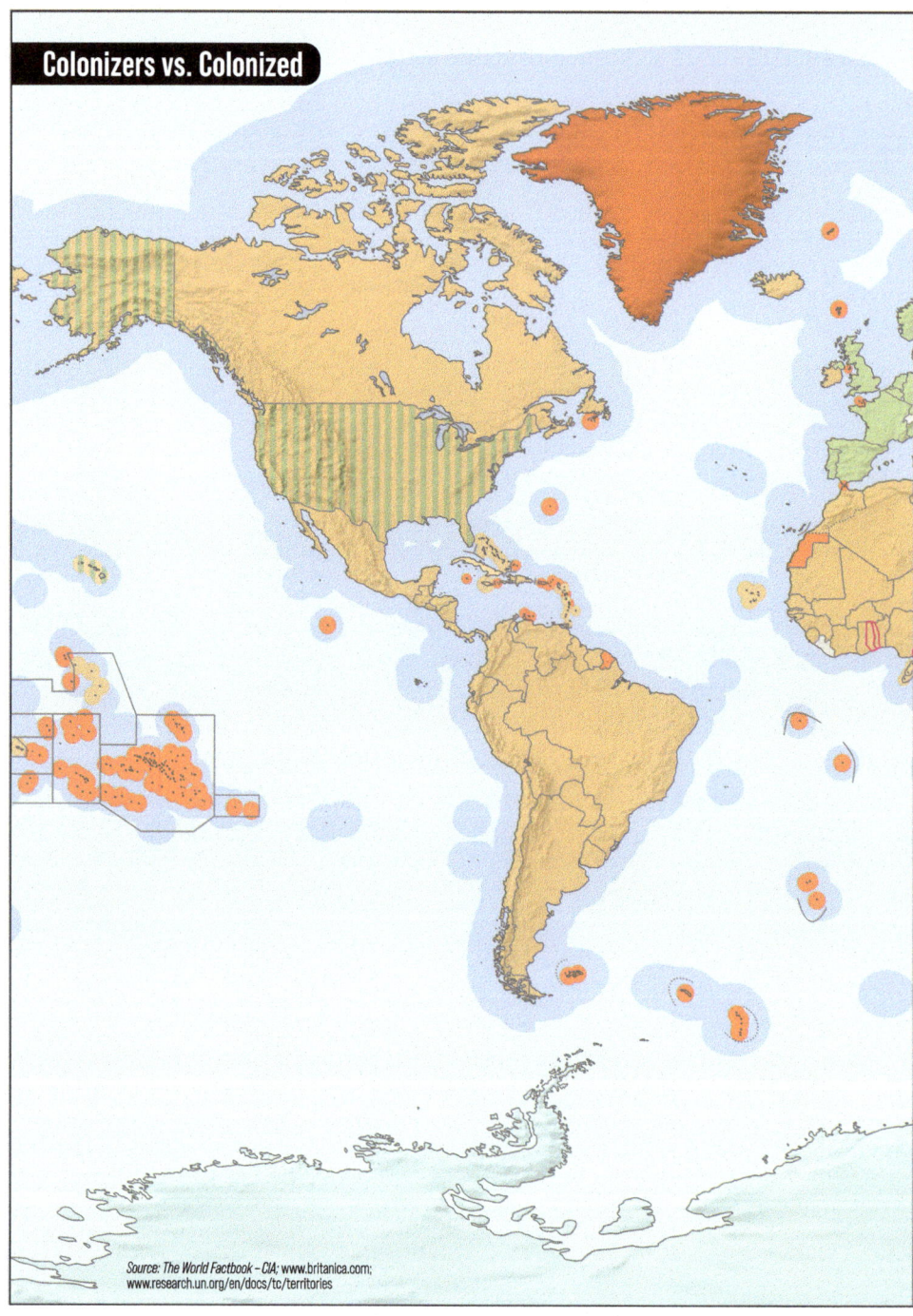

*Figure 7.12* The colonisers and the colonised

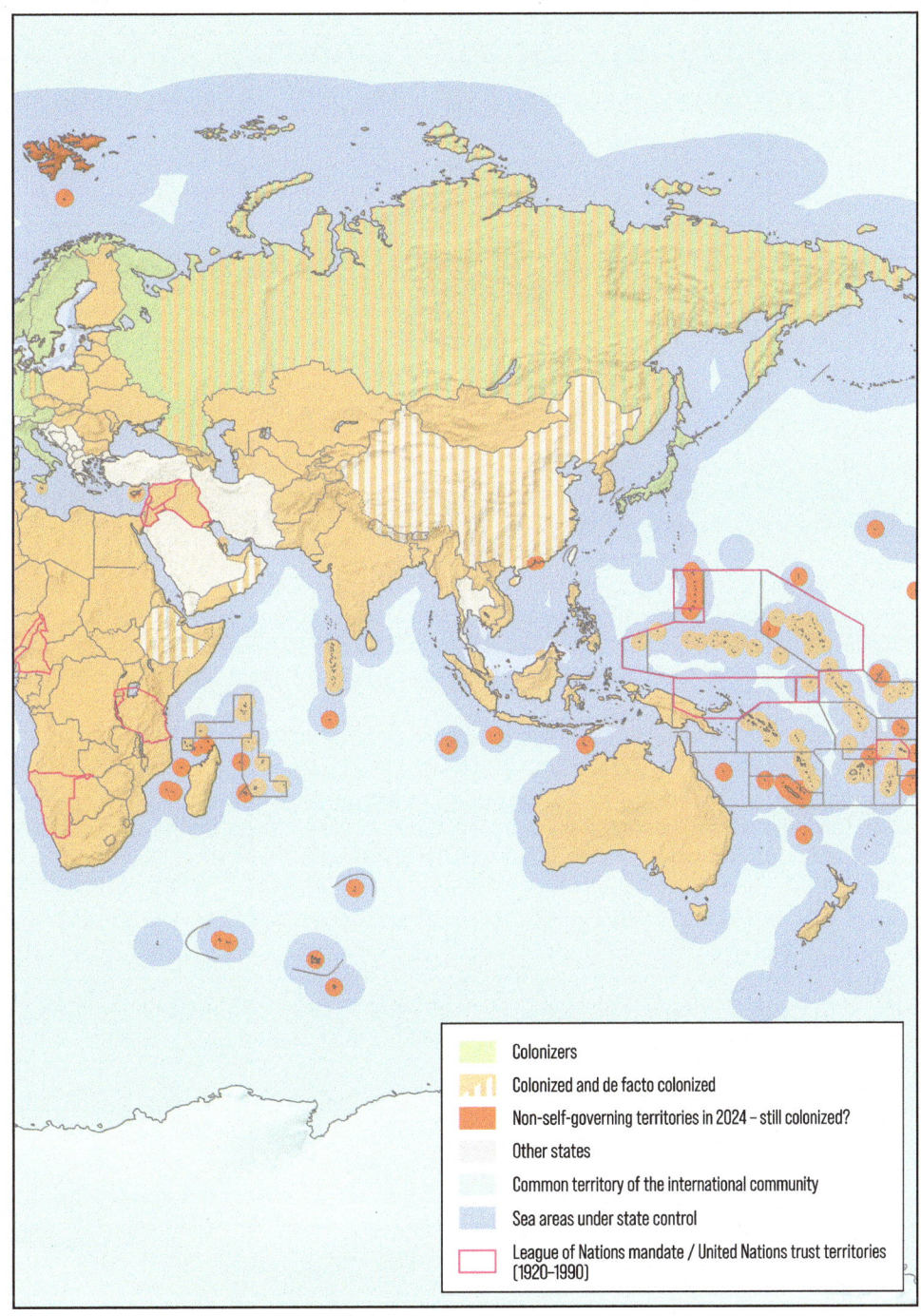

Colonizers

Colonized and de facto colonized

Non-self-governing territories in 2024 – still colonized?

Other states

Common territory of the international community

Sea areas under state control

League of Nations mandate / United Nations trust territories (1920–1990)

*Figure 7.12* (Continued)

# 8  Deconstructing the 20th and 21st centuries – the new Hundred Years' War

Norman Davies (1996) begins his description of the Hundred Years' War between England and France (1337–1453) by pointing out that it was neither officially declared nor uninterrupted. In short, he claims that 'Hundred Years' War' is merely a label used by historians to denote a period of chaos. When looking at the 'Long Twentieth Century' (1914–2024) through the lens of that mediaeval Anglo-French conflict, as succinctly characterised by Davies, it is difficult to avoid the similarities between the events of 500–600 years ago, as he describes them, and those we have been experiencing for over 100 years. The period of chaos and misery that began in 1914 was ushered in by the collapse of Europe's global hegemony (this statement of time coincidence does not involve a value judgement of the causes, practices, and effects of the domination of some European nations over others in Europe and beyond) and the birth and rivalry of two totalitarian regimes. Nor has this been an uninterrupted war – at least not a 'hot' one. Although certain rules were still observed in 1914, nobody declared a 'Hundred Years' War' at the time. It was generally thought that the war would be over in 3–4 months (Davies, 1996). Since then, conventions such as declarations of war have fallen into desuetude, as evinced by Germany in 1939 and Russia in 2022. The fundamental difference between the current and mediaeval (as described by Davies) situations is that there is no useful label for the period that began with a gunshot in Sarajevo on 28 June 1914.

In principle, there is a generally accepted narrative regarding the WWI (1914–1918), interwar (1918–1939), WWII (1939–1945), Cold War (1945/1946–1989/1991), and post– Cold War (after 1989/1991) tranches of the period from 1914 to 2024. However, not only are the dates delimiting these periods debatable, but the overall periodisation is open to challenge. The time limits covering the period 1914–2024 can be blurred and see a single indivisible whole in this period; one that requires a cumulative analysis:

> On the interpretational front, many years passed before historians began to ponder the unity of the 'European Civil War'. [. . .] Now, with the benefit of hindsight, it is increasingly clear that the successive conflicts formed part of one dynamic process: the two World Wars were separate acts of the same drama. Above all, the main contestants of the Second World War were created by the unfinished business of the First. By entering into military conflict in 1914, the European states unleashed the mayhem from which were born not one but two revolutionary movements – one of which was crushed in 1945, the other left to crumble in the dramatic events of 1989–1991.
>
> (Davies, 1996, p. 900)

DOI: 10.4324/9781003406440-8

As for the conclusion of WWI, Davies claims: 'Even then it was not decisive: the "Great Triangle" of military-political power blocs was not resolved until 1945, and in some respects not until 1991' (Davies, 1996, p. 901).

The sequence of events triggered by the assassination of Archduke Franz Ferdinand in Sarajevo on 28 June 1914 culminated in the July Crisis of 28 July–4 August and the outbreak of WWI (Nouschi, 2003). War in Europe might have been unavoidable, but only a tragic combination of happenstance and incompetence could have let this small spark detonate a chain reaction of events which set in motion a historical process that engulfed the next 100 years and cost tens of millions of lives. The signing of the Armistice at Compiègne on 11 November 1918 only signalled the end of WWI in Western Europe. The Bolshevik Revolution in Russia in 1917 triggered the following sequence of events in Eastern Europe: a partial decolonisation of the Russian empire began when nations under Russian rule strove for emancipation, the Russian Civil War (1917–1922) and the intervention of Allied forces (1918–1922), and an attempt to 'export' the communist revolution, which was brought to a halt by the Russian defeat in the Polish-Soviet War (1919–1921). WWI thus continued for another 4–5 years in the East.

The peace treaties that concluded WWI were unable to ensure a lasting peace. As WWII was to break out in CEE, it can be argued that this where the key to war and peace was to be found. As Davies puts it, 'the fault-line of the earthquake zone ran along Germany's eastern border' (Davies, 1996, p. 900). Whereas it was nationalist tensions in Austria-Hungary and the Balkans that triggered the sequence of events that culminated in WWI, the catalyst for WWII, apart from a historical revisionism grounded in a refusal to accept the outcome of WWI, was the national issue of the area bounded by the Adriatic, Baltic, and Black Seas. The mixture of ethnicities in this region prior to 1945 was so great and complex (Magocsi, 2019) as to foredoom any attempt to demarcate undisputed national borders in a world of rampant nationalisms. Moreover, the great powers were additionally laying the ground for future conflicts by ignoring national issues when drawing up the borders of the new CE European countries. While it is obviously not known whether observing the principle of national self-determination more closely when delimiting post-war borders would have preserved the peace, that desideratum was at least possible. Drawing up the borders of the countries that emerged after WWI, together with new borders for CEE that had existed before it began, was tantamount to loading a powder keg of claims, disputes, and conflicts – both between old and new states and among the latter. The numerous border conflicts associated with the German and Soviet revisionism that resulted from their dissatisfaction with the outcome of WWI led directly to a new war.

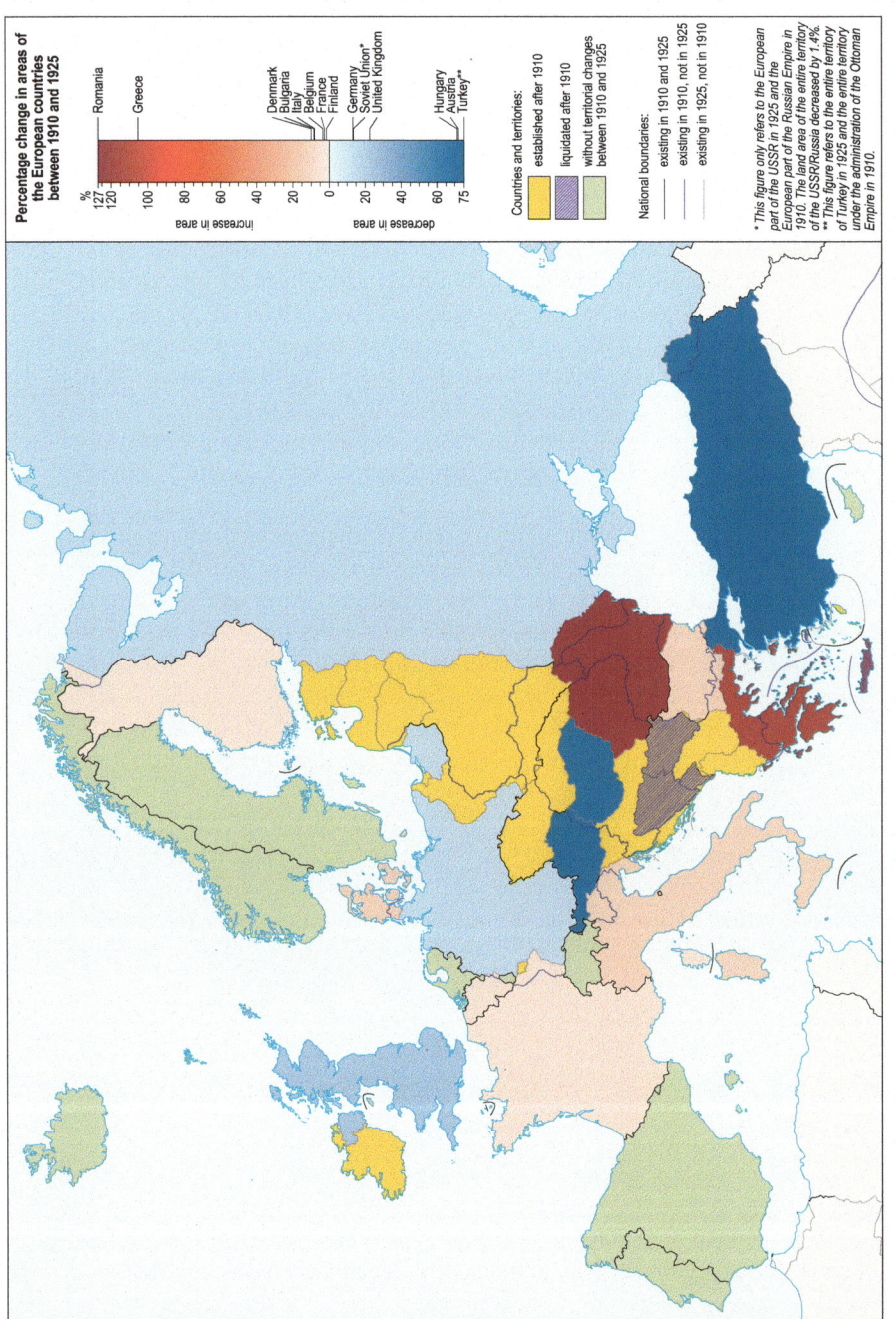

*Figure 8.1* Territorial changes in Europe: 1910–1925

*Source:* Own elaboration based on: *Annuaire Statistique International 1926* (1927) Genève: Société des Nations; *United nations demographic yearbook 1948* (1949) New York: UN DESA; *The statesman's year-book. Statistical and historical annual of the states of the world for the year* (1914, 1915, 1917, 1926, 1927, 1928) London: Macmillan and Co., Ltd.; *Rocznik statystyki Rzeczypospolitej Polskiej* (1925/1926, 1927, 1928, 1929, 1930) Warszawa: Statistics Poland; *The Rand McNally new library atlas map of Europe* 1:8,820,000 (1912) Chicago: Rand–MacNally & Co.; *Turkey in Europe & the Balkans* 1:2,150,000 (1910) London: Carmelite House, The London Geographical Institute; *A map of the countries between Constantinople and Calcutta including Turkey in Asia, Persia, Afghanistan & Turkestan* 1:6,969,600 (1912) London: Edward Stanford Ltd.; *Artaria's Eisenbahnkarte von Österreich-Hungarn* 1:1,400,000 (1911) Wien: Verlag von Ataria & Co.; *Atlas öfver Finland* (1910) Helsingforst: Sälskappet för Finlands Geografi; *Das Deutsche Reich in 4 Blättern* 1:1,500,000 (1911) Gotha: Justu Perthes; *Wielki atlas historyczny* (2002) Warszawa: Demart; *Omniatlas – interactive atlas of world History*, https://omniatlas.com.

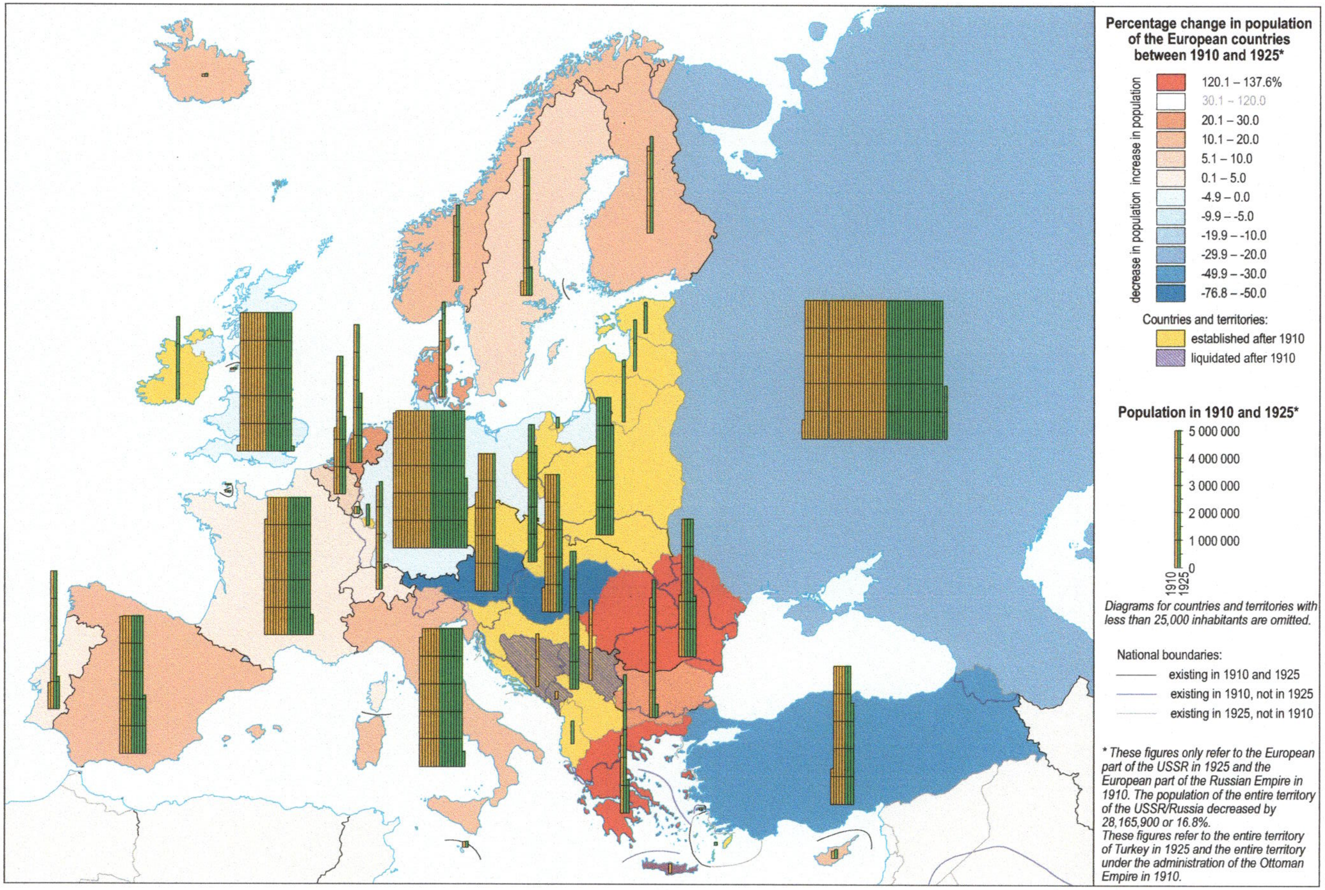

*Figure 8.2* Population changes in Europe: 1910–1925

*Source:* Own elaboration based on: *Annuaire Statistique International 1926* (1927) Genève: Société des Nations; *The statesman's year-book. Statistical and historical annual of the states of the world for the year* (1914, 1915, 1917, 1926, 1927, 1928) London: Macmillan and Co., Ltd.; *Rocznik statystyki Rzeczypospolitej Polskiej* (1925/1926, 1927, 1928, 1929, 1930) Warszawa: Statistics Poland; Mitchell, B. R. (1975) *European historical statistics, 1750–1970*. London: Macmillan Press Ltd.; Jefferson, M. (1914) 'Population estimates for the countries of the world from 1914 to 1920', *Bulletin of the American Geographical Society*, 46(6), pp. 401–413; *The Rand McNally new library atlas map of Europe 1:8,820,000* (1912) Chicago: Rand-MacNally & Co.; *Turkey in Europe & the Balkans 1:2,150,000* (1910) London: Carmelite House, The London Geographical Institute; *A map of the countries between Constantinople and Calcutta including Turkey in Asia, Persia, Afghanistan & Turkestan 1:6,969,600* (1912) London: Edward Stanford Ltd.; *Artaria's Eisenbahnkarte von Österreich-Hungarn 1:1,400,000* (1911) Wien: Verlag von Ataria & Co.; *Atlas öfver Finland* (1910) Helsingforst: Sälskappet för Finlands Geografi; *Das Deutsche Reich in 4 Blättern 1:1,500,000* (1911) Gotha: Justu Perthes; *Wielki atlas historyczny* (2002) Warszawa: Demart; *Omniatlas – Interactive Atlas of World History*, https://omniatlas.com.

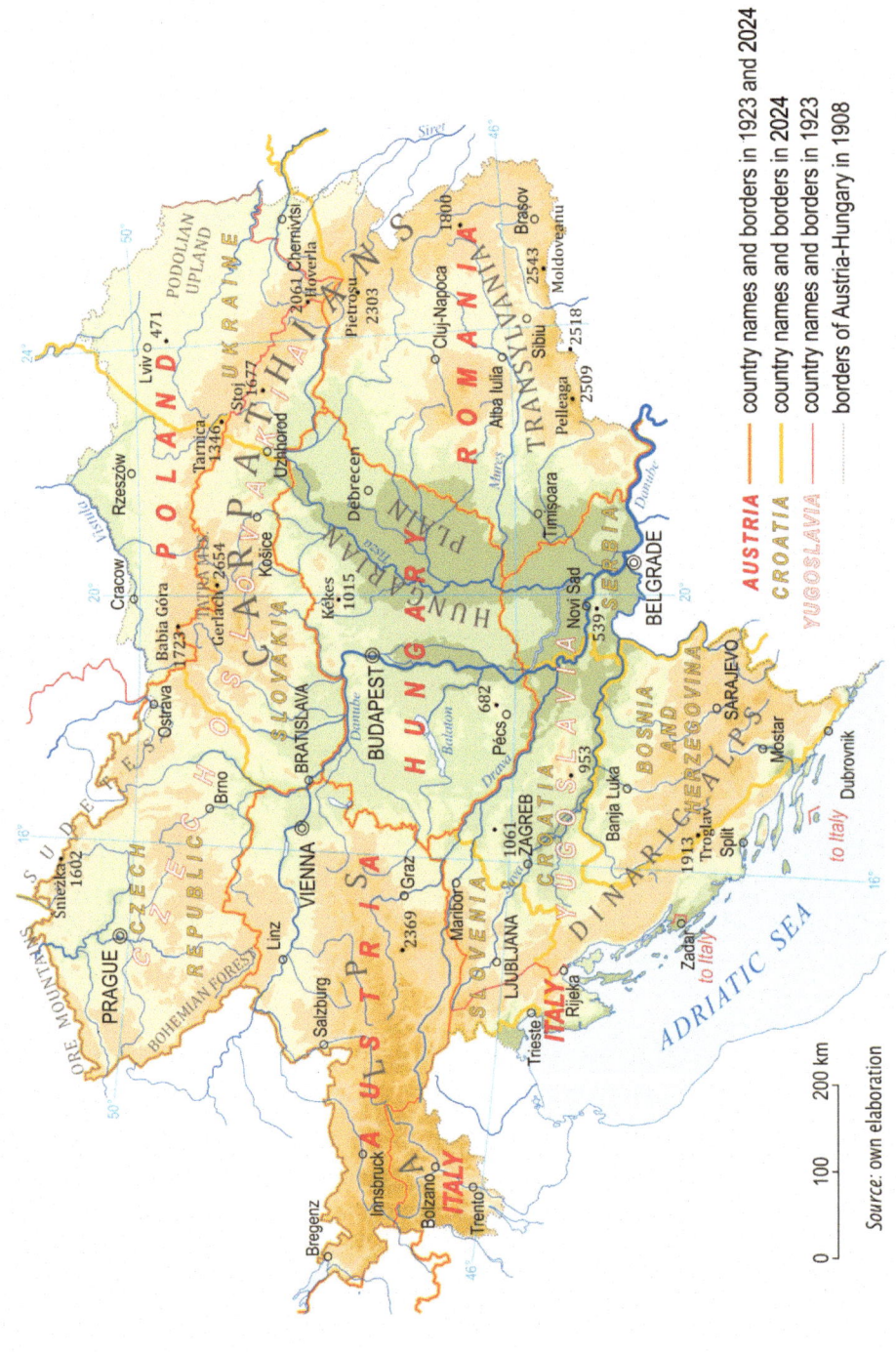

*Figure 8.3* Austro-Hungarian successor states

The most serious territorial conflict in CEE was the Polish-German dispute that arose directly from the peace resolutions of 1919–1921 and which led directly to war in 1939. The inequitable and punitive terms of the Treaty of Trianon (1920) towards the Kingdom of Hungary created another epicentre of grievance, distrust, and antagonism that enmeshed Hungary, Czechoslovakia, Romania, and Yugoslavia and set the country on a revisionist policy course that drove it into the sphere of influence of the Third Reich. The rise of aggressive nationalist ideas and movements was superimposed onto this state of affairs, effectively narrowing the scope for compromise. Hence, the odds of several ethnicities living together in peace in a given country and neighbouring countries with national minorities peacefully coexisting were considerably decreased. The multi-ethnic countries were doomed to fail. The ethnic 'knot' in CEE had been entangled by historical processes dating back hundreds of years, as evidenced by the ethnic maps of that period. Modern concepts of nationalism, chauvinism, social Darwinism, and natural determinism, however, pulled the separate threads together so tightly that it became impossible to disentangle them, and the knot itself became unacceptable to many people.

Last but not least, WWI had given birth to the totalitarian systems of National Socialism (Nazism) in Germany and Soviet Communism in Russia. Neither accepted the other, and each rejected the existing international order, claiming new territories in the name of (respectively) 'living space' and 'world revolution'. In CEE, they were separated geographically by a minimal distance and politically by weak or very weak countries that were usually at loggerheads. The region was therefore a preordained crumple zone between two totalitarian powers on a collision course. The untrammelled (but opposing) expansionism immanent to both made confrontation inevitable. 'It was a duel which in the hands of totalitarian revolutionaries was destined to become a fight to the death' (Davies, 1996, p. 900). Despite the annihilation of Nazism in 1945, however, Russian propaganda took up the cudgels against it once again in 2022. In 1921, American writer and businessman Pierrepont Noyes aptly and presciently likened Europe to an automobile with faulty brakes rolling down a steep slope. Disaster was a surety, although the form it would take was not yet known (Zdziechowski, 1993, p. 363).

A strong and united system of alliances acting as a vise to press Germany from the east would have helped ensure the stability of Europe after WWI, but for a variety of reasons, France and the UK were unable to assemble one (Gawron, 2005). France's most powerful allies in CEE, viz. Poland and Czechoslovakia, were mutually hostile in both word and deed in 1918–1939 (Gawron, 2005, p. 48), as evidenced by the Czechoslovak military attack of 1919 and Poland's retaliatory participation in carving up Czechoslovakia in 1938. Meanwhile, simple political geography would appear to dictate that the two countries work together, but unfortunately, they were not prisoners of their geographical location and borders in this respect. They were both allied with France, they had a strong cultural affinity (they were in the Latin cultural sphere and their majority populations were of West Slavic descent), they had a common (albeit occasionally troublesome) history, and they shared a long border (measuring 984 km [Kubiczek *et al.*, 2003, p. 257] and comprising 17.7% of Poland's borders and 32% of Czechoslovakia's in 1937). Most significantly, perhaps, they had many common geopolitical determinants, the main one being contiguity with Germany. However, they viewed the Soviet Union differently, although this is also explicable with reference to the political map. Czechoslovakia did not have a border with the USSR and it did not lie in the CE European Lowlands, i.e., the crucial conflict zone between East and West in Europe. Poland not only had a

1,412-km border with the USSR (Kubiczek *et al.*, 2003, pp. 257–258) but also sharply divided this conflict zone where it was almost at its narrowest. The Soviets additionally undermined Poland's viability as a state. Finally, Polish-Russian relations were heavily burdened by history: Polish statehood had been threatened and vassalized from the 17th century before being completely erased by Russia in the 18th century, and every Polish attempt to regain independence after 1794–1795 was forcefully suppressed. Despite that, however, the geopolitical position of the two countries was surprisingly similar considering their contiguity with Germany and one other hostile country (Poland–the USSR; Czechoslovakia–Hungary).

A cursory glance at the political and ethnic maps of CEE reveals several factors prodding Poland and Czechoslovakia to cooperate. First, their geographical shapes were unfavourable. Czechoslovakia was long and narrow and tapered from west to east-southeast, while Poland had peninsulas that jutted into neighbouring countries and wide territorial bays. Their shapes were mostly unfavourable, however, with respect to Germany. Obviously, close cooperation would not have removed the challenges and threats stemming from their shapes, which each eventually had to face alone in the late 1930s (Czechoslovakia was progressively dismantled as a state by the Third Reich in 1938–1939 following Germany's annexation of Austria [Der Anschluss]; Poland was invaded by Germany from the north, west, and south in 1939), but it would definitely have mitigated them, as the political map makes clear. The narrow Pomeranian corridor that gave Poland access to the Baltic (its window on the world) but which also separated Berlin from the German exclave of East Prussia was not the only problem that Germany posed for Poland. Greater Poland was surrounded by German territory from the north, west, and south, and German Silesia pushed far into the southeast, driving a wedge between Czechoslovakia and Poland. The upshot was that all of Poland west of a line joining Cieszyn with Suwałki, i.e., roughly one-fourth the country, was in Germany's clutches. After the partition of Czechoslovakia and the vassalisation of the newly independent Slovakia in 1938–1939 by Germany, Poland's situation only deteriorated, as German pincers also pressed Poland from the south, holding already roughly 40–50% the country (the most developed part). The situation facing landlocked Czechoslovakia was comparable to that of Poland prior to the Anschluss (1938), but incomparably worse thereafter. It was then, however, similar to the Polish situation after Czechoslovakia had been liquidated in 1939. Before 1938, the Czech part of Czechoslovakia cut deep into German territory in the west and north. After Austria's incorporation into Germany, however, the entire 'hull' of Czechoslovakia west of a line joining Ostrava with Bratislava, an area roughly coextensive with the present Czech Republic, was encircled by Germany. Bohemia and Moravia were now territorially connected with Slovakia by a narrow 'Czechoslovakian isthmus' some 150-km long squeezed by German pincers.

Territorial shape was associated with access to the sea, which had considerable significance for sovereignty, security, and economic development. On gaining independence, following the decisions taken at the Paris Conference in 1919–1920, Poland had access to the Baltic, but only at the end of a long (150 km), narrow (25–100 km) neck of land without a seagoing port. Danzig (Gdańsk) was supposed to serve Poland's needs, but it was outside the country's borders (it had been part of Poland in 1466–1793), having been made a Free City, ethnically dominated by Germans and resolutely ill-disposed towards Poland. Making Gdańsk a Free City under the auspices of the League of Nations was a compromise that neither Poland nor Germany found satisfactory and that created a permanent flashpoint in mutual relations. Poland tried to partially rectify this adverse situation by building a new seaport in Gdynia from scratch and linking it to the rest of the country, but this did not solve the crux of the problem, viz. the difficulty of defending a narrow neck of land that was encircled by Germany and that was to become one of the official reasons for the German invasion of Poland in 1939. Czechoslovakia was at an even greater disadvantage by virtue of having no maritime access whatsoever. This not only hampered the country economically but, most importantly, also completely vitiated its alliance with France (Czechoslovakia also had a mutual assistance treaty with the USSR, but as it did not have a land border with that country, any Soviet assistance would have had to pass through Poland, which was completely impossible).

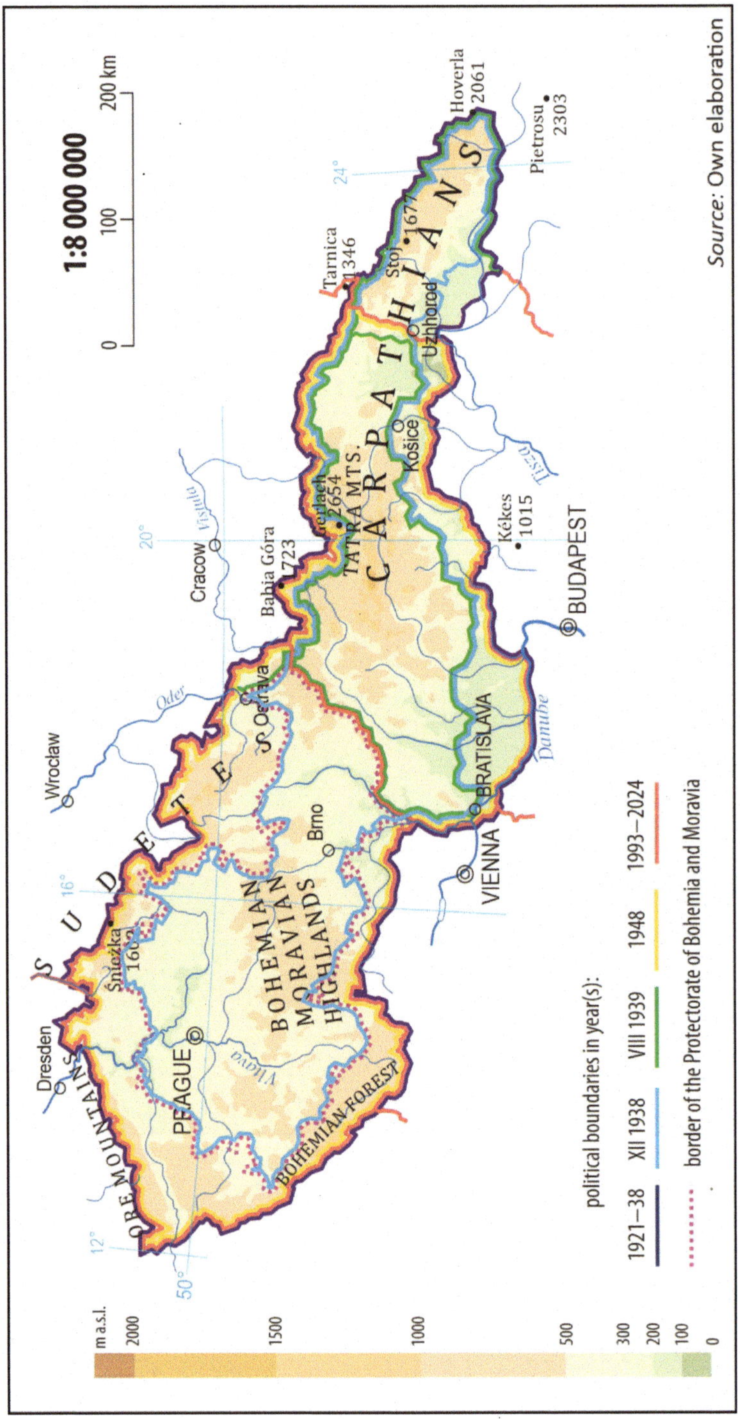

*Figure 8.4* Development of the borders of Czechoslovakia: 1921–1992

*Figure 8.5* Czechoslovakia – borders and territory: 1937–1992

*Source:* Own elaboration

Second, a highly complex, irregular territorial shape comes with many drawbacks, but security need not be overly burdensome for an insular country or one surrounded by allied, or at least non-aggressive, neighbours. What counts is the context that creates the international environment. Poland and Czechoslovakia, however, had few neutral, let alone friendly, neighbours. Poland's only friendly neighbours in 1918–1939 were Latvia and Romania. These were weak countries that had short and extremely peripheral borders with Poland (106–109 km and 347–349 km, respectively [Kubiczek *et al.*, 2003, pp. 257–258], together comprising 8.5% of its terrestrial borders). Warsaw was separated from Berlin, Moscow, Kaunas, Prague, and Gdańsk by hostility or aversion, i.e., with its other neighbours, which together accounted for 91.5% of its land borders (Poland gained a short border with friendly Hungary at the expense of Czechoslovakia in 1939). Moreover, Germany and the USSR impugned the very existence of the Polish state, while relations with Lithuania were strained by an intractable dispute over Vilnius and environs. Vilnius was the historical capital of Lithuania but had a majority Polish population. The sticking points in Poland's relations with Czechoslovakia were incidental compared with those in its relations with Germany and the USSR, and their most serious territorial conflict, i.e., over Cieszyn Silesia, could have been resolved by compromise as early as 1918. However, this was never done (Solarz, 2022, p. 80). Apart from having an unfriendly Poland to its north, Czechoslovakia bordered Austria, Germany, and an increasingly revisionist Hungary. Its sole ally (and only against Hungary) was Romania, at its far eastern end, and its border with that country constituted less than 5% of its borders. During the Central European crisis of 1938–1939, Czechoslovakia lost even this friendly border when Hungary annexed and occupied Carpathian Ruthenia in March 1939. Faced with German and Soviet aggression, Poland partially exploited the goodwill of Romania and the now contiguous (1939) Hungary, with which it had a 277-km border (Kubiczek *et al.*, 2003, p. 258). However, this did not alter its fate.

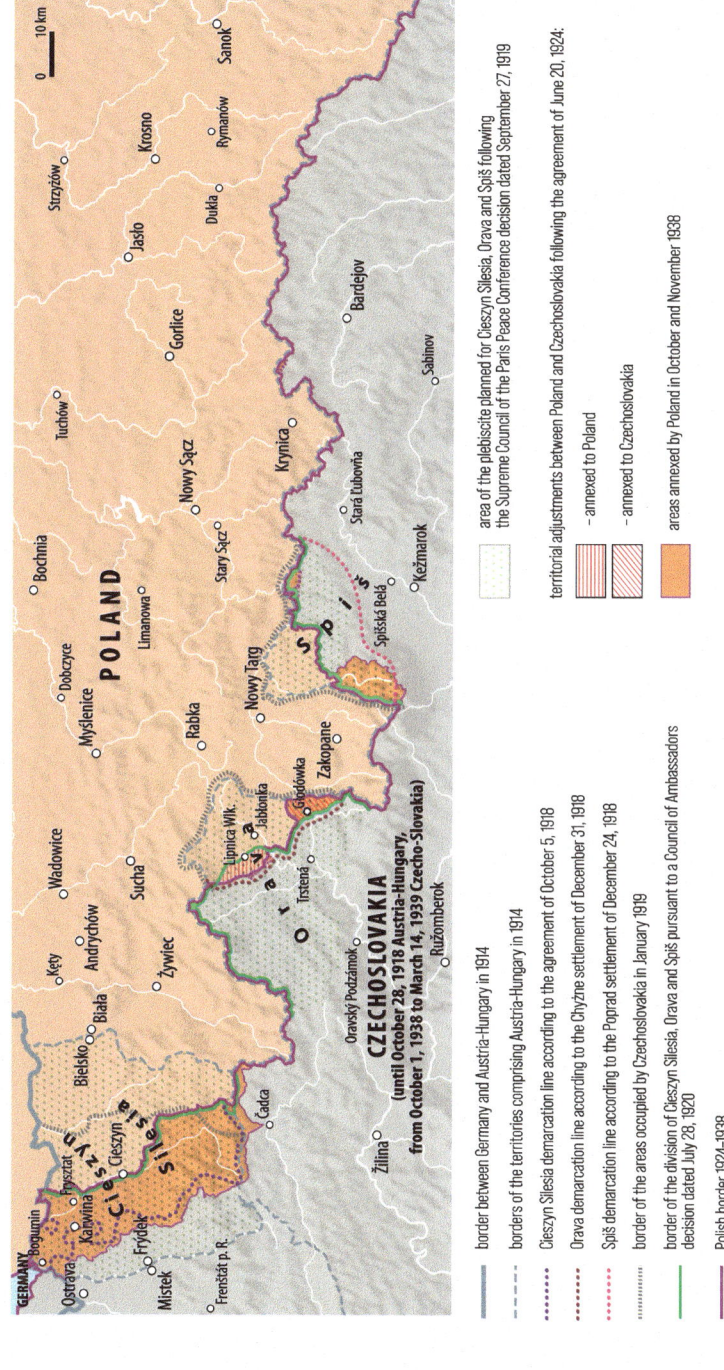

*Figure 8.6* Disputes between Poland and Czechoslovakia during the interwar years

*Figure 8.7* Disputes between Poland and Czechoslovakia: 1939–1958

Third, German minorities posed a significant problem for both countries. Almost the entire ethnic German population in the 'Czech part' of Czechoslovakia was concentrated in a compact strip along the border with Germany and Austria. Germans comprised 22.3% of the Czechoslovakian population in 1930, but in Bohemia and Moravia, the figure was 29.2%. Not only that, but 4.8% of the population was made up of Hungarians hostile to Prague (Hungarians comprised 17.6% of the population of the 'Slovak part' of the country, where they mostly lived near the Hungarian border). In 1930, these 3.2 million Germans not only constituted Czechoslovakia's largest ethnic minority but they even outnumbered Slovaks (the country's third largest ethnic group; Slovaks made up 15.8% of the population; Czechs were the country's largest ethnicity, accounting for 51.1% of the population) (Magocsi, 2019, pp. 141–143). By contrast, Germans were only the fourth largest minority in Poland, after Ukrainians, Jews, and Belarusians. In 1931, they accounted for barely 2.3–3% of the population, but still numbered 741,000, which was almost a quarter the size of the Czechoslovakian German community (Poland was more than twice the size of Czechoslovakia and had more than double its population in the interwar period). As in Czechoslovakia, Polish Germans mostly lived near the German border, specifically in western and northern Greater Poland and in Pomerania, the latter being the territory that granted Poland its politically and economically crucial maritime access (Magocsi, 2019, pp. 105, 131; Solarz, 2022, p. 96). The flawed treaties imposed by the Paris Peace Conference (1919–1920), European nationalist ideas and movements, the weakness and shortsightedness of the Western powers, and finally, 'minor' Central European quarrels meant, as Pierrepont Noyes foresaw, that another war was merely a matter of time. Not surprisingly, the flashpoint was where Germany bordered Czechoslovakia and Poland.

The revisionist German, Hungarian, and Soviet states were creatures of the treaties signed at the Paris Peace Conference and soon afterwards. None of them had the slightest interest in maintaining the status quo *post bellum*. Italy soon joined them, while the United States withdrew into isolationism. For their part, the guarantors of the Versailles order, France and the UK, suicidally began to destabilise it (e.g., by the 1925 Locarno Treaties and the 1938 Munich Agreement). Germany became especially active in destabilising the Versailles order once Hitler became chancellor in 1933 by withdrawing from the League of Nations (1934); reincorporating the Saar Basin (1935); remilitarising the Rhineland and building up its armed forces (1936); annexing Austria and adjacent regions of Czechoslovakia (1938); and seizing or subordinating the rest of the 'hull' of Czechoslovakia, annexing Lithuania's Klaipėda Region, colluding with the USSR to divide CEE into spheres of influence (the Ribbentrop-Molotov Pact), and, finally, invading Poland (1939). Seeing one revisionist country prevail with no opposition or significant costs emboldened others. Hungary clearly emulated Germany's triumphs. It participated in carving up Czechoslovakia by annexing southern and eastern Slovakia and Carpathian Ruthenia (1938–1939), and it took northern Transylvania from Romania (1940) and Bačka from Yugoslavia (1941). Italy initially focused on conquests in Africa, but seized Albania in 1939. The Ribbentrop-Molotov Pact (23 August 1939) and subsequent Soviet expansion were also underpinned by the victories of, and lack of sanctions against, the Third Reich. The USSR carved up Poland (with Germany and Slovakia); invaded Finland (1939); and incorporated Lithuania, Latvia, Estonia, and part of Romania (1940). As a result, not only did a Soviet-German border appear (1939) and lengthen (1940), but two aggressive revisionist regimes with opposing orientations faced off there. The inevitable war between them commenced on 22 June 1941. Europe erupted in open warfare when

Germany and the USSR invaded Poland in 1939, but the annexation of Austria and the dismemberment of Czechoslovakia in 1938 had been clear steps towards it. Even before that, however, Italy had invaded Ethiopia in 1935 and Japan had invaded Manchuria (China) in 1931 and the rest of China in 1937. The interwar period was therefore not only characterised by the inherent instability of the Versailles order, brought about by the flaws in the peace treaties it imposed, the dissatisfaction of certain countries with the outcome of WWI, and the lack of a workable international security system, but this interlude between the two successive wars was shorter than is generally supposed. Globally, it lasted less than 10 years, and at most 17 (1922–1939) in Europe. The second world war was therefore born of the first.

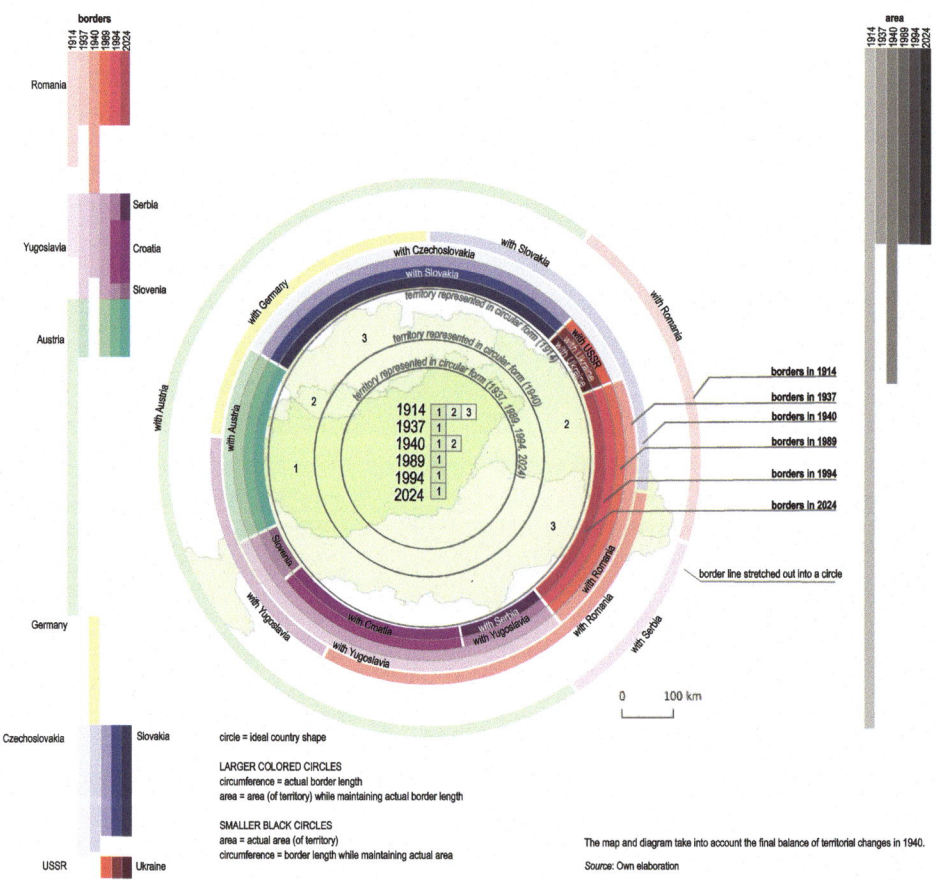

*Figure 8.8* Hungary – borders and territory: 1914–2024

*Figure 8.9* Romania – borders and territory: 1914–2024

The map and diagram take into account the final balance of territorial changes in 1940 and 1941.

*Source:* Own elaboration

circle = ideal country shape

LARGER COLORED CIRCLES
circumference = actual border length
area = area (of territory) while maintaining actual border length

SMALLER BLACK CIRCLES
area = actual area (of territory)
circumference = border length while maintaining actual area

*Figure 8.10* Development of the borders of Hungary: 1914–2024

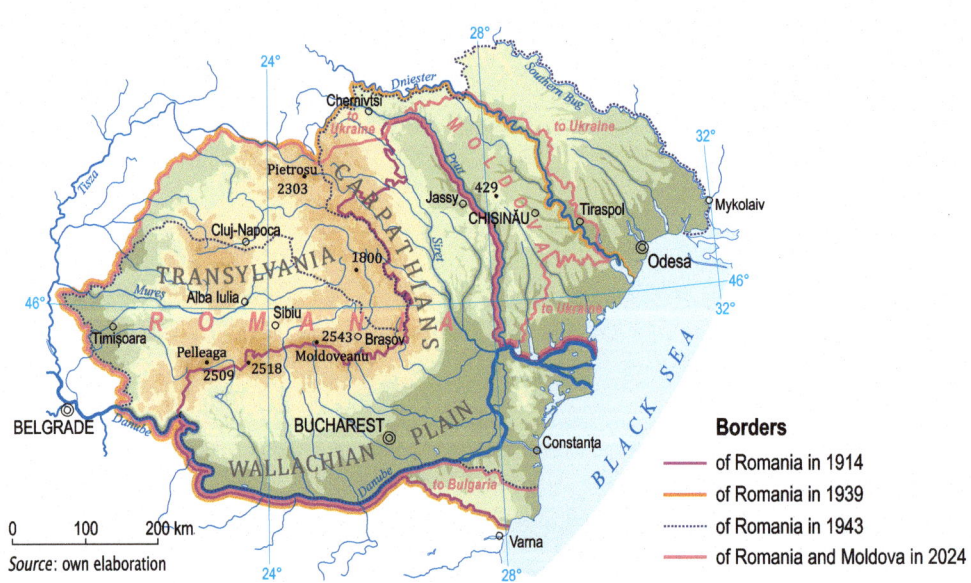

*Figure 8.11* Development of the borders of Romania: 1914–2024

The territorial outcomes of WWII did not satisfy everyone any more than did those of the first. The greatest beneficiary of the territorial changes in Europe was the USSR, which was larger than it had been in 1937. Not only did it keep the fruits of its 1939–1941 alliance with Hitler (which seemed permanent during the Cold War), as set out in the Ribbentrop-Molotov Pact, but it built a tightly controlled and closely supervised sphere of influence behind a closed 'iron curtain'. This outcome was nevertheless unsatisfactory for the USSR from the standpoint of its ambitions. Despite everything, the territorial losses of Germany, Poland, and Czechoslovakia came to be accepted over time (80 years after WWII, border revisionists were a fringe element in the first two countries, while Czechoslovakia was dissolved in 1993). All things considered, despite the two latter countries being deeply dependent on the USSR until 1989, the border changes were at least neutral for Czechoslovakia, and actually beneficial to Poland. The significant and painful Polish territorial losses in the underdeveloped multinational East, including major urban centres such as Vilnius and Lviv and the forced population resettlements, were more than offset by the acquisition of more developed territories in the West and the resolution of the ethnic tensions that had seriously weakened interwar Poland (preserving the pre-war ethnic tapestry seriously destabilised the country, especially in the wake of the Polish-Ukrainian conflict). Germany had lost another war, and once more, part of the price it paid was territorial. This time, however, the penalties were more severe. After WWI, Germany lost 13% of the territory it had held in Europe in 1910, but now it lost an additional 24% of what it had held in 1937. Germany was dissolved and four occupation zones (American, British, French, and Soviet) were installed in its stead. Austria was separated from Germany, thereby nullifying the Anschluss of 1938, and was likewise divided into four occupation zones. A united Austria reappeared as early as 1955, whereas Germany, having been divided into the pro-West Federal Republic of Germany (FRG) and the pro-Soviet German Democratic Republic (GDR) in 1949, did not manage to reunify until 1990. In 1945, Germany lost the northern part of East Prussia, along with Königsberg (Kaliningrad), to the USSR and all the land east of the Lusatian Neisse and the Oder (Stettin [Szczecin] and the land west of the river were an added bonus), along with the southern part of East Prussia, to Poland. It was stripped of all territory east of the Lusatian Neisse and the Oder that it had arduously acquired since the Middle Ages, including Pomerania, which Brandenburg-Prussia had gained from the Thirty Years' War in 1648 and the Great Northern War in 1720, the land ceded by Austria following the Silesian Wars (1740–1763), and the territory taken from Poland and, later, the Grand Duchy of Warsaw following the Partitions of Poland (1772, 1793, and 1795) and the Napoleonic Wars (1815). Despite forfeiting 45.6% of its 1939 territory to the USSR and suffering much greater territorial losses than Germany (the largest in Europe, despite formally being among the winners), Poland was the main beneficiary of Germany's territorial losses (32.4% of the territory of contemporary Poland was obtained at the expense of Germany and 0.6% the Free City of Gdańsk) and was second only to the USSR when it came to territorial acquisitions in Europe (although the country was reduced in size by 20% after the war) (Solarz, 2014a, 2022). This is probably one of the main reasons why Poland has not experienced anything like Hungary's 'Trianon syndrome'. Apart from losing the peripheral and impoverished Carpathian Ruthenia to the USSR, Czechoslovakia was reconstituted as it had been before the war, although it gained a small area south of Bratislava at the expense of Hungary in 1947. Germans were expelled from Poland and Czechoslovakia when their borders were changed or their pre-war borders restored. Combined with additional waves of post-war migrations, this led to a complete change in their ethic mixes (in Poland, however, remembering not only the mass displacements and subsequent emigration but also taking into account the significant shift of the country westward and the population effects of the Holocaust organised and carried out by the Germans) and solved their German minority problems. Whereas ethnic minorities had comprised 31.1–36.2% of Poland's population in 1931, with Germans alone accounting for 2.3–3%, the respective figures in 2011 were 3.12% and 0.19% (Solarz, 2022, p. 96; Magocsi, 2019, p. 131). These decisions on border changes and population exchanges in CEE, which were essentially made by Stalin with the consent or acquiescence of Roosevelt and Churchill, effectively reversed the course that history had taken since the Middle Ages (Piskorski, 2001, p. 125) and, paradoxically, in terms of the outlook for peace and stability in Europe (which Stalin had almost certainly not intended) cut through the ethnic Gordian knot in many places in CEE. For its part, Poland, in addition to being in thrall to Soviet imperialism, as well as both a territorial victim and beneficiary of it, was a potential addressee of German territorial claims. It therefore had no interest (which is still true today) in challenging post-war border decisions. By giving Poland a third of its current territory at the expense of Germany (Solarz, 2022, p. 59), Stalin was hoping (ultimately in vain in the long historical perspective) that Poland would be permanently dependent on, and subordinated to, the Soviet Union out of fear of German revisionism. This was because the USSR was the only reliable guarantor of its territorial acquisitions in the West. Over time, however, the German political mainstream has come to terms with its permanent territorial reduction (the 1950 Treaty of Zgorzelec, the 1970 Treaty of Warsaw, the 1990 Treaty on the Final Settlement with Respect to Germany, the 1990 German-Polish Border Treaty), but it was unable to undo the division of Germany into the FRG and the DRG until 1990. Five Ds (demilitarisation, denazification, democratisation, decentralisation, decartelisation) were imposed on Germany after 1945. Some of these restrictions were waived over time (demilitarisation) and some (denazification) were applied too superficially and inconsistently, but they were generally effective, and Germany was made non-revisionist – at least for the time being. While the Nord Stream pipeline is seen as a reincarnation of the spirit of Rapallo, which had been so harmful to the European order in the past (pursuant to the 1922 Rapallo Treaty Germany and Soviet Russia secretly agreed to cooperate militarily in violation of the Versailles Treaty), the resolutions following WWII put an end to the cycle of provoking conflicts within the framework of a new Hundred Years' War at least so far as Germany was concerned.

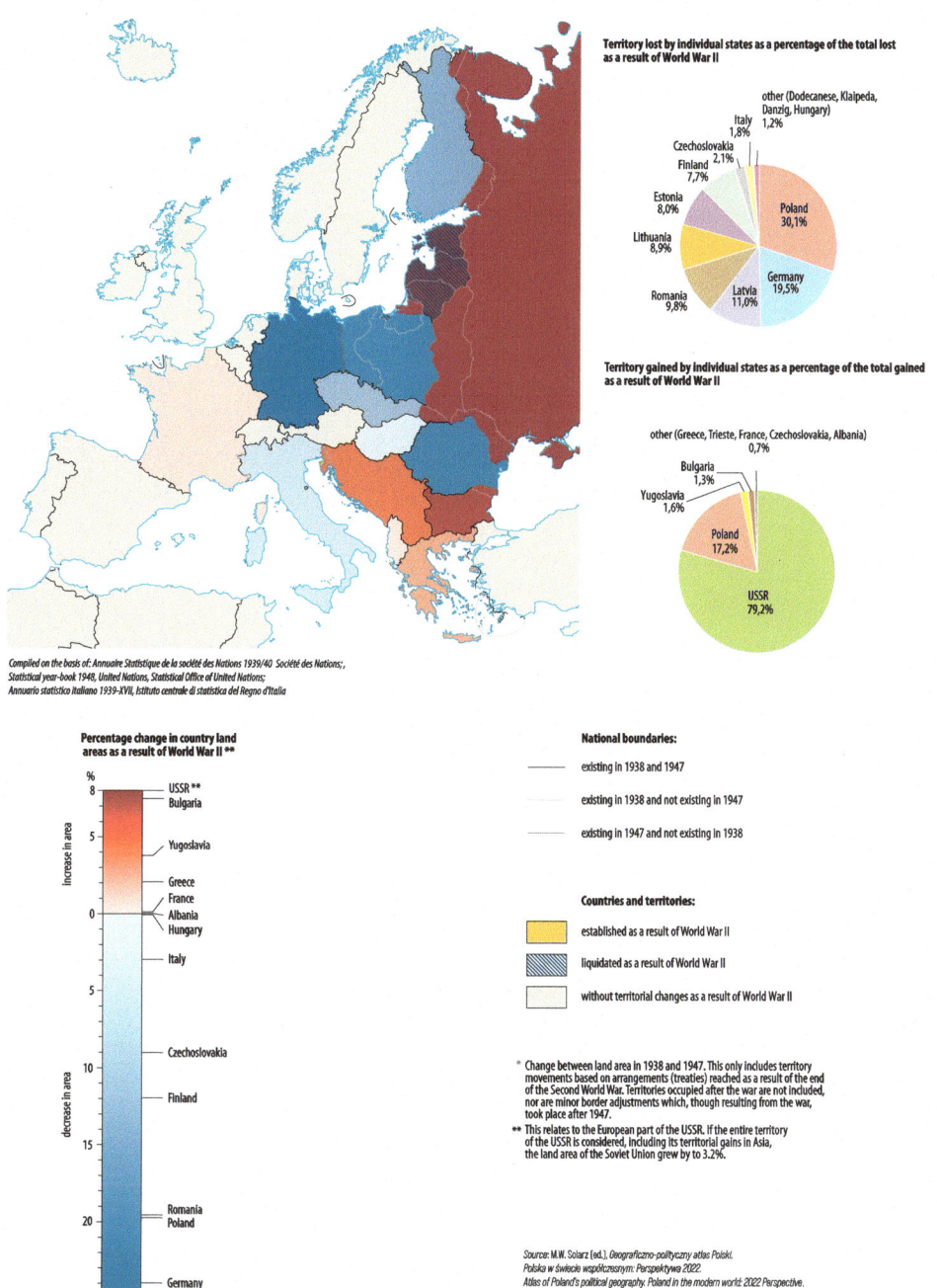

*Figure 8.12* Territorial changes in Europe following World War II

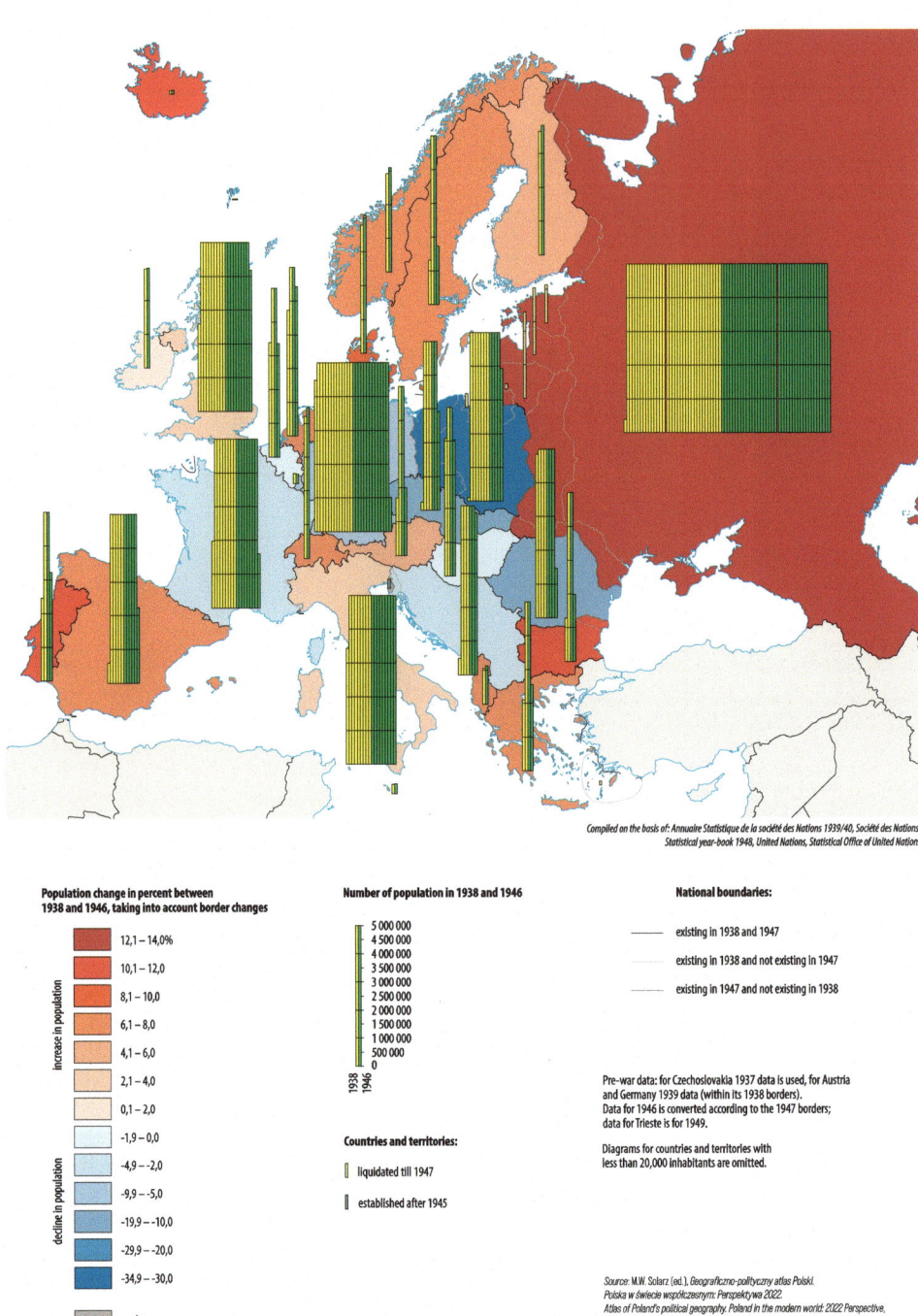

Compiled on the basis of: Annuaire Statistique de la société des Nations 1939/40, Société des Nations;
Statistical year-book 1948, United Nations, Statistical Office of United Nations

**Population change in percent between 1938 and 1946, taking into account border changes**

increase in population

| | |
|---|---|
| | 12,1 – 14,0% |
| | 10,1 – 12,0 |
| | 8,1 – 10,0 |
| | 6,1 – 8,0 |
| | 4,1 – 6,0 |
| | 2,1 – 4,0 |
| | 0,1 – 2,0 |

decline in population

| | |
|---|---|
| | -1,9 – 0,0 |
| | -4,9 – -2,0 |
| | -9,9 – -5,0 |
| | -19,9 – -10,0 |
| | -29,9 – -20,0 |
| | -34,9 – -30,0 |
| | no data |

**Number of population in 1938 and 1946**

5 000 000
4 500 000
4 000 000
3 500 000
3 000 000
2 500 000
2 000 000
1 500 000
1 000 000
500 000
0

1938  1946

**Countries and territories:**

liquidated till 1947

established after 1945

**National boundaries:**

existing in 1938 and 1947

existing in 1938 and not existing in 1947

existing in 1947 and not existing in 1938

Pre-war data: for Czechoslovakia 1937 data is used, for Austria and Germany 1939 data (within its 1938 borders).
Data for 1946 is converted according to the 1947 borders; data for Trieste is for 1949.

Diagrams for countries and territories with less than 20,000 inhabitants are omitted.

Source: M.W. Solarz (ed.), Geograficzno-polityczny atlas Polski.
Polska w świecie współczesnym: Perspektywa 2022.
Atlas of Poland's political geography. Poland in the modern world: 2022 Perspective,
Warszawa 2022. With the kind permission of the Office of the "Niepodległa" Program.

*Figure 8.13* Population changes in Europe following World War II

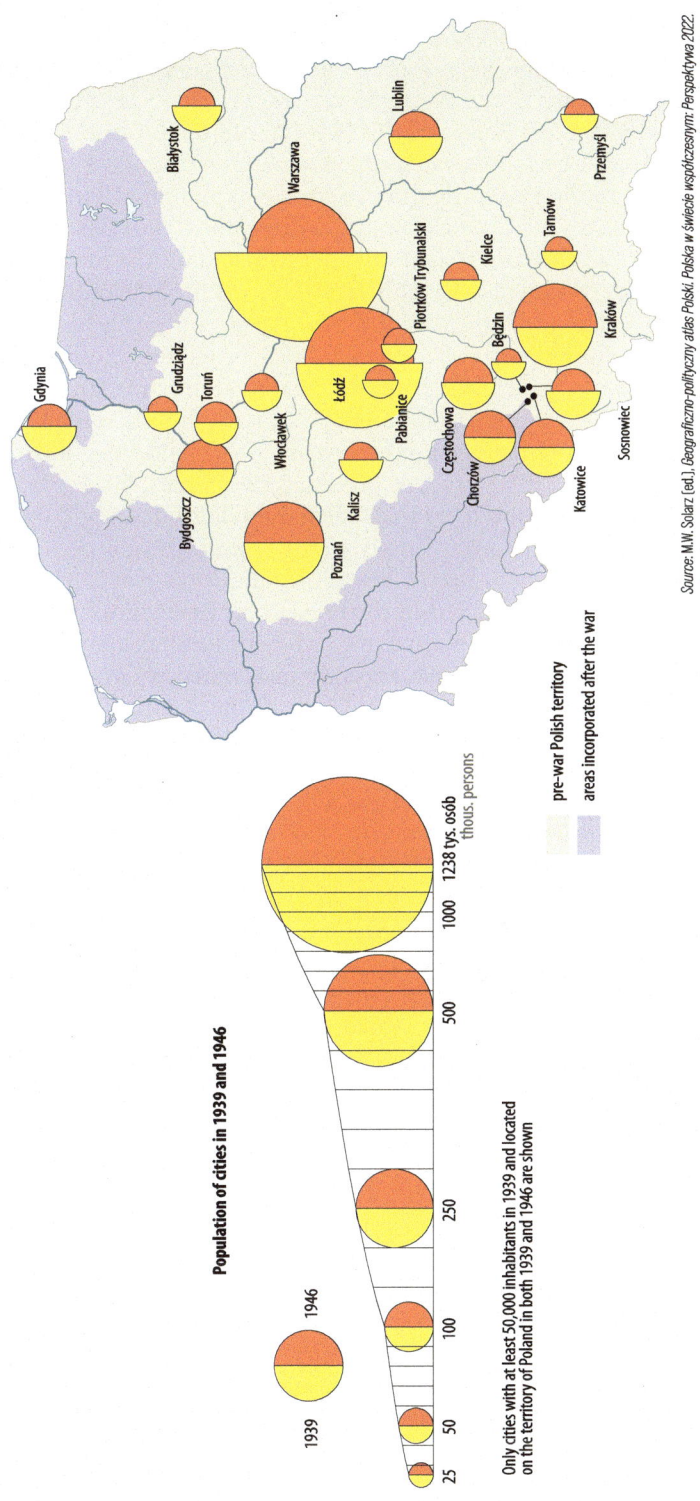

Source: M.W. Solarz [ed.], *Geograficzno-polityczny atlas Polski: Polska w świecie współczesnym: Perspektywa 2022. Atlas of Poland's political geography. Poland in the modern world: 2022 Perspective*, Warszawa 2022. With the kind permission of the Office of the "Niepodległa" Program.

*Figure 8.14* Population changes in Polish cities following World War II

Over the short term, however, the outcome of WWII was decidedly unfavourable from the standpoint of CEE. The Red Army had advanced as far as the Elbe, imposing communist totalitarianism in place of Nazism. In so doing, the USSR had at least partially reversed the results of its defeat in the Polish-Soviet War (1919–1921). As Churchill put it in a speech he gave in Fulton, Missouri, in 1946: 'From Stettin in the Baltic to Trieste in the Adriatic, an iron curtain has descended across the Continent' (Davies, 1996, p. 1065). The countries to the east of that line, with the exception of Lithuania, Latvia, and Estonia, and to the north of Yugoslavia and Greece retained their place on the global political map, but forfeited their sovereignty to Red Russia until the turn of the 1990s. Until the end of the 1980s, these countries showed no respect for human rights, including political rights and civil liberties, and they fell increasingly further behind in development due to the absence of market mechanisms in their economies. Some Eastern Bloc societies, notably Poland, strongly rebelled against this situation. In Poland, there was an armed anti-communist underground (1940s and 1950s; it is officially recognised that the last member of the anti-communist resistance movement was killed in combat in 1963), worker revolts (1956, 1970, 1976), student protests (1968), and, finally, 'Solidarność' [Solidarity], a nationwide mass movement that called for independence and social change (1980–1981) (officially, Solidarność was simply a trade union not subject to the communist authorities) and which was suppressed following a military coup in 1981. In Hungary, opposition to Soviet subjugation erupted in the Hungarian Revolution (October 1956), which was bloodily crushed by the Soviet Army, while in Czechoslovakia, the 'Prague Spring' liberalisation reforms (1968) were abruptly terminated when other communist countries invaded that year. The precipitous collapse of the Soviet sphere of influence, from the Elbe to the Bug and the Black Sea, in 1989–1991 (this putative superpower, along with its sphere of influence, ceased to exist in the space of three years), shows that CEE never willingly accepted the order that the USSR imposed after WWII and which it maintained by force. This regional chapter of WWII (itself a consequence of Hitler's pact with Stalin in 1939) was not permanently closed until the adverse effects of the Ribbentrop-Molotov Pact on the sovereignty of the countries between Germany and the USSR were finally annulled almost 50 years after Germany's defeat in 1945. Poland possibly gives the best illustration of this. Poland was reborn as a sovereign country in 1918 following the defeat of Germany, Austria-Hungary, and Russia in WWI. Its reappearance on the political map of Europe was not accepted by either Germany or Russia. Poland's major military objectives in 1939 were to maintain its independence and its territorial integrity. The two-pronged invasion in 1939 (Germany on 1 September and the USSR on 17 September) quashed both. Being part of the anti-Nazi alliance from day one (Poland stood alone against the Third Reich on 1–3 September 1939; France and the UK formally entered the war on the side of Poland on 3 September, *de iure* marking the beginning of a coalition war) did not mean that any of the military objectives of 1939 would be achieved. In 1945, having been at war with the Third Reich longer than any other allied country, Poland regained its place on the map, but the USSR had stripped it of vast tracts of territory and deprived it of its independence, only restoring it with the end of the Cold War in 1989. National independence and a place on the political map, which had been achieved simultaneously in 1918, required two steps after WWII: Poland reappeared on the map in 1945 following Germany's defeat in the war, and full sovereignty was restored in 1989–1991 as a result of the USSR's Cold War failure and collapse. From Poland's perspective, WWII was the result of its being reborn after WWI in defiance of the wishes of Germany and Russia and reversing the consequences of Hitler's 1939 pact with Stalin in 1945 first and then in 1989–1991 was the combined result of WWII and the Cold War. For Poland, then, the events of 1914–1989/1991 were a linear sequence of causation.

*Figure 8.15* Development of the borders of Poland and the country's partition divisions: 1922–2024

*Figure 8.16*  Poland – borders and territory: 1922–2024

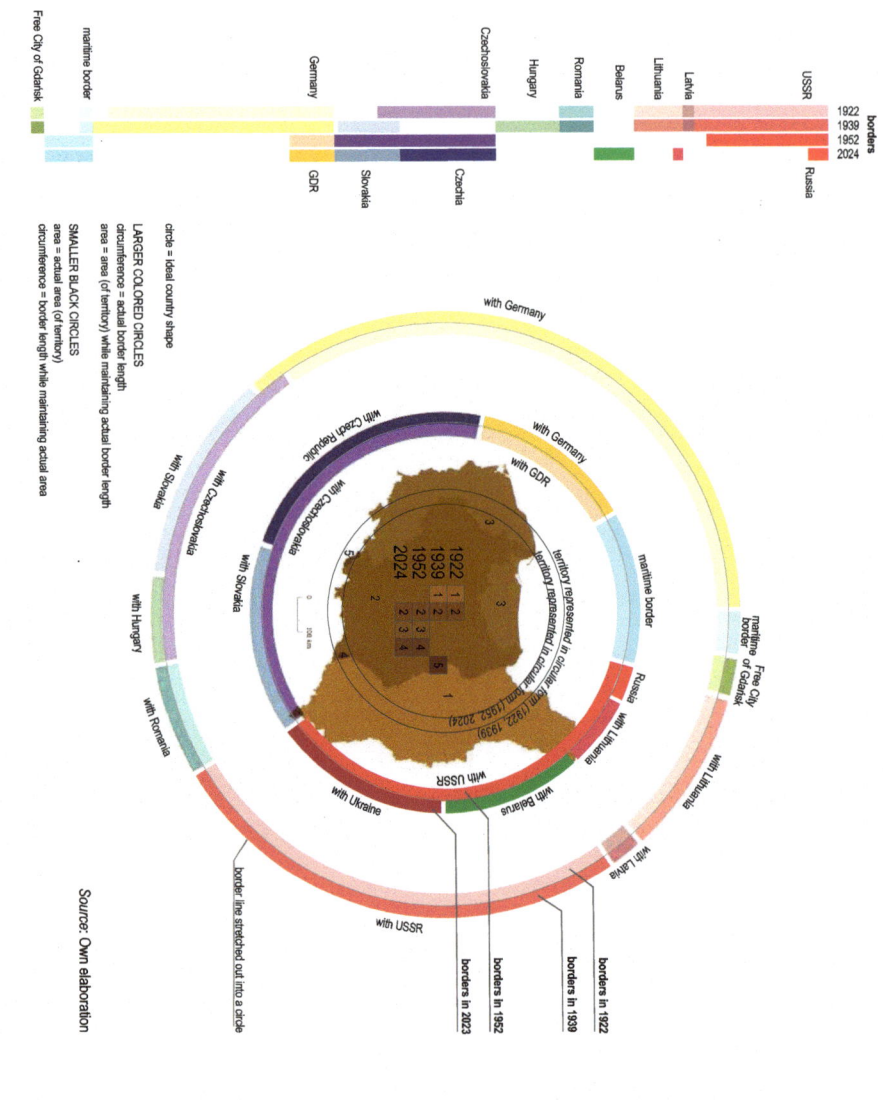

*Source:* Own elaboration

Thus, the new status quo after 1945 – neither solicited nor desired in CEE – had both a territorial and a political dimension. This is not only evident in the case of Poland. Hungary, occupied by the Red Army, lost all the territory it had gained at the expense of Czechoslovakia, Romania, and Yugoslavia after 1937 through its cooperation with Nazi Germany. Its policy of territorial revisionism had therefore been a very costly road to nowhere. The balance sheet for Germany – occupied, split, and territorially reduced – is the same. However, another aggressor, viz. the USSR, extended its borders up to 300 km westward in a 2,000-km strip stretching from Finland to Romania after WWII. Lithuania, Latvia, and Estonia, first annexed by the USSR in 1940, now remained within its borders until 1990–1991. In addition to these three countries, Finland, Germany, Czechoslovakia, Romania, and, most of all, Poland, lost territory to the USSR. The first four countries only lost relatively or, in fact, small peripheral areas. The USSR's policy of aggression and imperialism therefore paid off. This did not bode well for the future.

Russia gives an impression of the continuity of the historical process similar to that conveyed by Poland, albeit over a longer period; one that spans, at a minimum, 1914–2024. At the turn of the 1990s, Russia suffered its greatest geopolitical catastrophe since losing the Battle of the Kalka River in 1223, and the empire's borders receded to where they had been at the turn of the 18th century. The USSR, which was merely imperial Russia rebadged, quietly broke down and fell apart. There were no major battles or casualties. From a purely human standpoint, this was definitely a positive development. However, having their empire disintegrate in this manner did not induce the Russians to reflect on their responsibility for the turn that world history, and their own, had taken. Consequently, no reckoning was made, no guilty party was punished, and no necessary changes were made. In 1991, however, it seemed that the curtain had fallen on the final act of the drama that began with reign of the House of Romanov, which had ended the Time of Troubles (1598–1613). The Polish-Lithuanian Commonwealth reached its farthest eastward extent in 1619, and its army had occupied Moscow in 1610–1612. A Polish-Lithuanian-Russian union was even mooted at the time, but in the end, nothing came of it, and the 17th century was marked by two more Polish-Russian wars (the war of 1609–1618 was followed by those of 1632–1634 and 1654–1667). However, the Polish-Lithuanian Commonwealth (Kingdom of Poland and Grand Duchy of Lithuania until 1569) permanently lost the strategic initiative it had held in the east for 300 years as early as 1686. Symbolically, Russia then seized it with the Treaty of Perpetual Peace (1686) and held it for the next 300 years (Mieroszewski, 1974, pp. 11–12). This period appears to have ended in 1991. After 1697, the Polish-Lithuanian Commonwealth became a Russian vassal before progressively vanishing in three partitions (1772, 1793, 1795) and in its defeat in the defensive war with Russia in 1792 and the Kościuszko Uprising against Russia and Prussia in 1794. The Russian border was eventually based on the Nemen, Bug, and Zbruch Rivers in 1795. Russia also broke the power of rival Sweden during the Great Northern War (1700–1721), and for the first time ever, its army advanced deep into Germany in the Seven Years' War (1756–1763). Following the revolutionary and Napoleonic wars, Russia extended its borders even farther westward, past the Vistula Line, and into the heart of CEE, co-determining the political and territorial configuration of Europe. It sustained territorial losses after WWI with the rebirth of Poland and Lithuania and the creation of Finland, Latvia, and Estonia. Its western border thus temporarily retracted until 1939: in the north, it was roughly where it had been prior to the Battle of Poltava (1709), and in the centre and the south, it was where it had been after the Second Partition of Poland (1793). Russian discontent at having lost territory as a result of WWI and having its World Bolshevik Revolution (1917) checked at the Battle of Warsaw (1920) and the Battle of the Neman River (1920) led to the USSR's short-lived agreement with the Third Reich in 1939, which it consummated in 1939–1941, to split CEE into Soviet and German spheres of influence. Germany's failed invasion of the USSR (1941–1945) enabled the Russian empire to push its borders even farther west than they had been in 1914 or 1939 (to be precise, the borders of its sphere of influence, not its own) – right into the heart of Germany. This, in combination with Soviet global manoeuvring, provoked another worldwide conflict, viz. the Cold War with the United States and its allies. The geopolitical catastrophe wrought by its Cold War defeat (stamped in 1989–1991), however, did not induce Russia to engage in any self-reflection or persuade it to reject the imperialist political paradigm. This might have been because the specifics of the Cold War and its conclusion spared Russia the sobering shock of total defeat. As it was not a major confrontation, the Cold War did not end in military defeat for the USSR, as WWI had for the German Empire and WWII for the Third Reich. No conference was held to impose peace conditions, similar to the provisions of the Treaty of Versailles (1919) and the Potsdam Agreement (1945) towards Germany, on the losing side. There were no trials. Instead, the North Atlantic Treaty Organization (NATO) and the Warsaw countries met in Paris to solemnly declare on 19 November 1990 (Joint Declaration of the Twenty-two Member States of NATO and the Warsaw Treaty Organization) that they were no longer enemies and to extend the hand of friendship to each other (Solarz, 2020, p. 9). As it had done in 1945, Russia evaded responsibility for starting a worldwide conflict. The USSR's defeat and the ensuing deep crisis of the 1990s, combined with a lack of reflection on Russia's role in destabilising the world order and an absence of penalties for doing so, was the perfect setting for the country to re-create itself as a new Weimar Republic destined to morph into an aggressive state, modelled on the Third Reich, determined to overturn the new political and territorial status quo.

The causal link between the two world wars has long since been found. Although there were many causes for WWII, dissension over the treaties and arrangements of 1919–1921 was one of the main ones. The next steps towards overturning the status quo put in place after WWI and entering into the more devastating WWII were the remilitarisation of the Rhineland (1936), the annexation of Austria (1938), the Munich Agreement (1938), partitioning the remainder of Czechoslovakia (1939), the Ribbentrop-Molotov Pact (1939), and finally, the German-Soviet invasion of Poland (1939). German impunity after 1933 demoralised other countries, especially the USSR, which decided to also take advantage of the Western policy of appeasement. Not only did WWI create new international problems with its contested territorial layouts, but it failed to settle the great power rivalry that was its root cause. Germany and Russia might have been defeated (the former by the Western alliance; the latter by Germany, Poland, and itself), but their ambitions were not curbed. To use sports parlance, WWII can be defined as a rematch or extra time. However, Germany suffered an even greater defeat in 1945. Not only did it temporarily cease to exist (until 1949), but it was split into two separate countries (1949–1990). Moreover, it lost all of the territory that it had annexed since 1938, and even some that it had held in 1937. It was rebuilt in accordance with the Five Ds of the Potsdam Agreement, although there are several reservations regarding this last point. First, it was not true in every respect: in some respects, it was not true at all, while in others, it was only partly or belatedly true. SS General Heinz Reinefarth, who lived a normal life in the FRG after the war, even sitting in the Schleswig-Holstein Landtag (state parliament), despite having murdered tens of thousands of civilians in the Warsaw Uprising in 1944 – a war crime for which he was

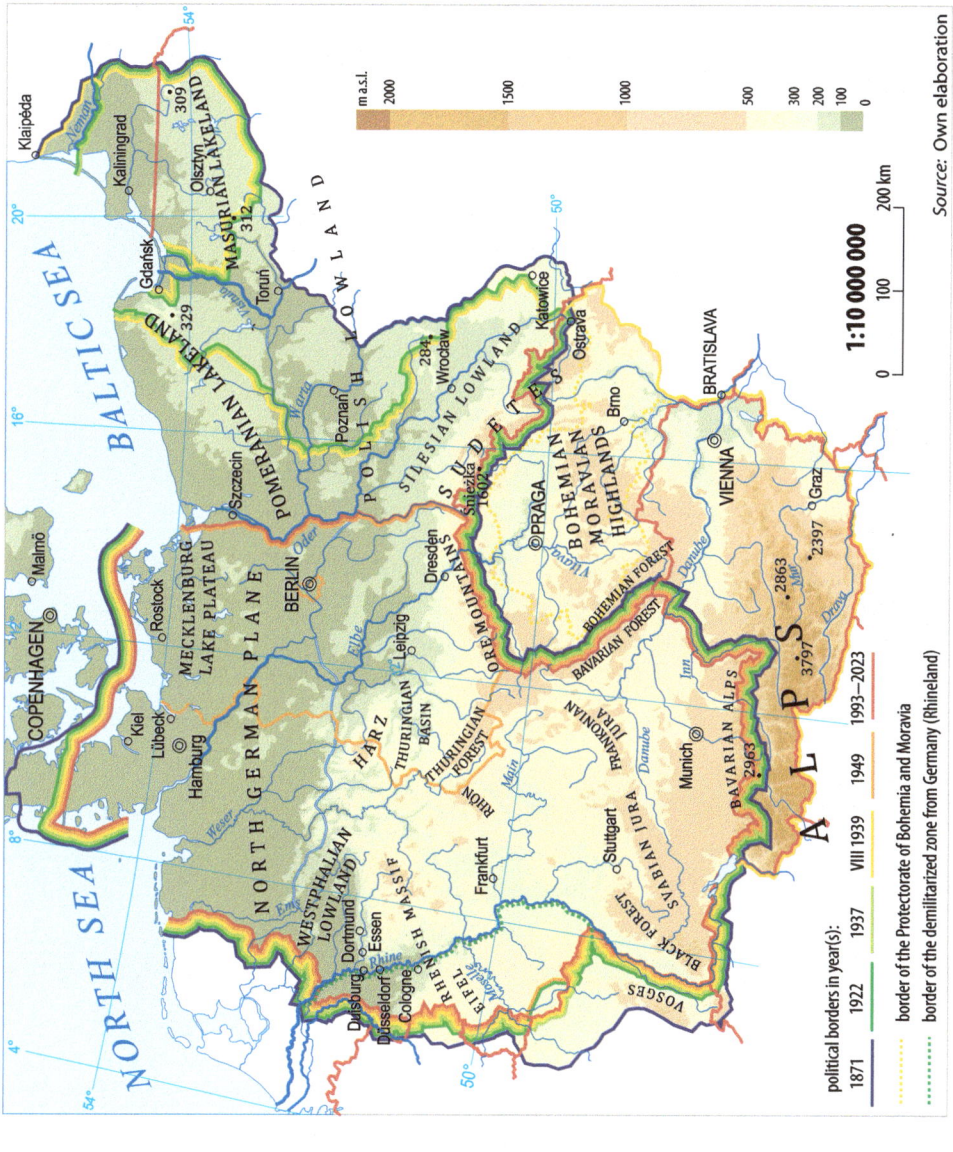

*Figure 8.17* Development of the borders of Germany: 1871–2024

*Figure 8.18* Germany – borders and territory: 1914–2024

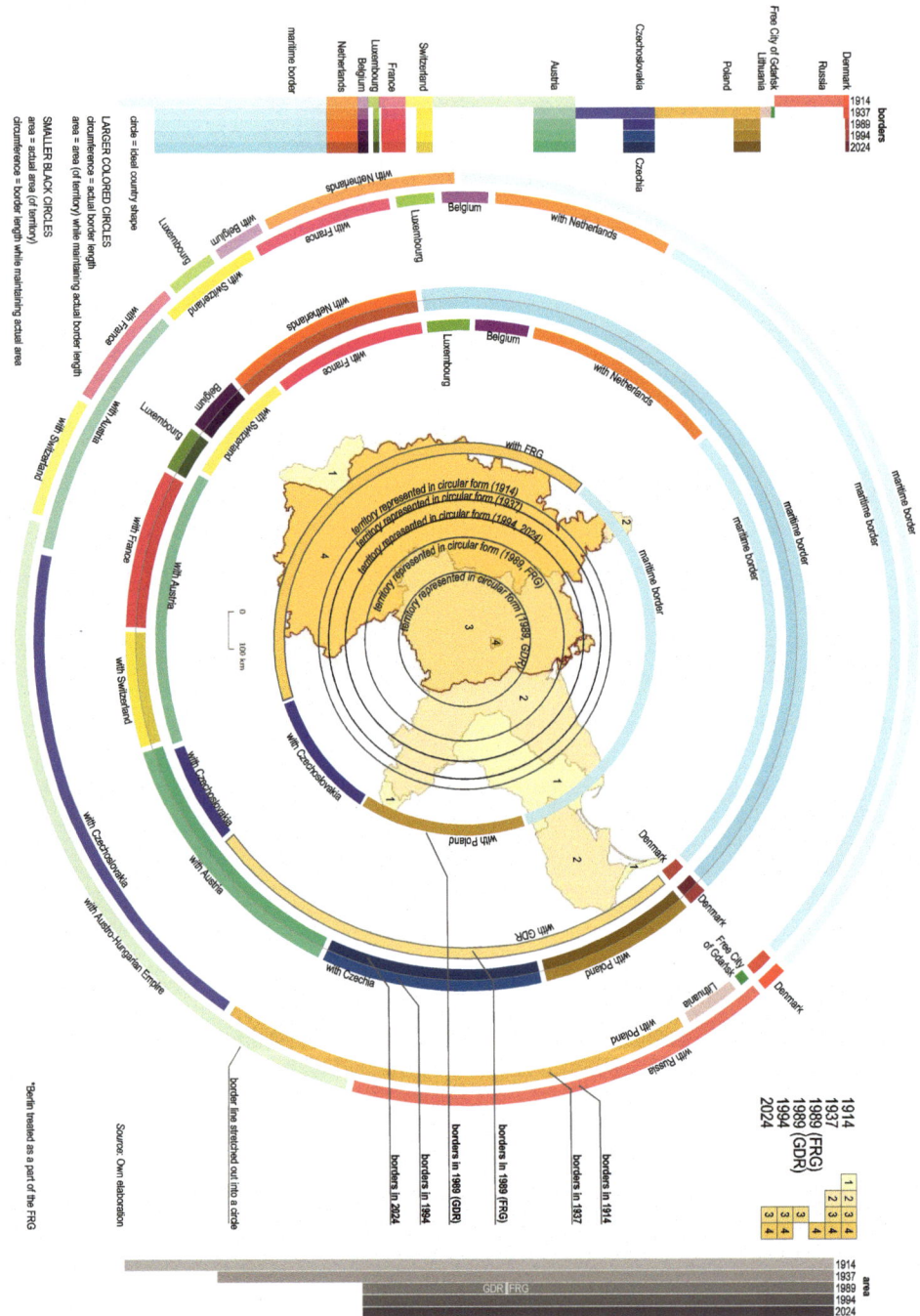

never punished – is an illustrative case in point regarding denazification. Moreover, both Germanys were members of military blocs from 1955, and there was no democratisation of the GDR until a few months prior to German reunification in 1990. Nevertheless, Germany's 1945 defeat was total and constituted a starting point for creating another Germany. However, this was not the case with the USSR, which had overtly collaborated with Nazi Germany in destroying the order established by the Versailles and Riga peace treaties by invading Poland and Finland (1939) and seizing part of Romania and annexing Lithuania, Latvia, and Estonia (1940). By the end of 1940, Germany had invaded 8 countries, whereas the USSR had 'only' invaded 6 (the ratio was 10:6 on the eve of Germany's invasion of the USSR). Not only has the USSR never borne any responsibility for its complicity in starting WWII and committing other aggressive acts, along with all the cruelty, deportations, destruction, and plunder that went with them (e.g., the Katyn Massacre [1940] in which it murdered almost 22,000 Polish prisoners of war [POWs] and civilians), it was their chief beneficiary. Its joint culpability with Nazi Germany for bringing about WWII was rewarded with new territories and a vast sphere of influence that extended into the heart of Germany. From the Soviet standpoint, its aggressive policies proved to be beneficial and profitable, despite the German invasion (1941) and the horrendous cruelty and massive losses that ensued. It therefore had every incentive to continue them, which it did in its global rivalry with the United States (1945–1991).

However, positing another Thirty Years' War, from the shots in Sarajevo (1914) to the Potsdam Agreement (1945), with two world wars along the way (Davies, 1996; Solarz, 2012), does not suffice to name and explain the chain of cause and effect that carved the political map of the world throughout the 20th century and into the 21st. Like its immediate predecessor (WWII), the Cold War was a great worldwide conflict (a de facto WWIII) that followed on from a preceding world war and ended without any conclusive settlements (or at least not everyone was satisfied). It is depicted as a confrontation between two political, military, economic, cultural, etc., superpowers with competing political and development models. Its basic feature was constant escalation: 'horizontal' – from Europe to the rest of the world (Kennedy, 1995, p. 371) – and 'vertical' – into new areas of life. Any description of the Cold War has at its core the two global superpowers; their ambitions to rule the world; the large political and military blocs they led; the sheer scope of their rivalry, which even extended into outer space; their numerous peripheral wars (usually conducted by proxies); and the great political processes they sustained and escalated and which significantly altered the political map of the world, as witnessed by e.g., decolonisation and European integration. Paradoxically, the only reason it never became 'hot' was the destructive capability of nuclear weapons (Solarz, 2020; Davies, 1996).

The Cold War, however, was also a strange war: peace–non-peace and war–non-war, at the same time a kind of shadow theatre, performance, illusion. François Furet (1996, p. 9) notes that 'the Soviet regime fled the theatre of history in panic, although it had arrived in fanfare'. This counterpoint between great power and house of cards calls the reality of the conflict between the United States and the USSR as two superpowers into question. Genuine bipolarity implies equal capabilities, but as these two countries were in no sense equal rivals, or realistic alternatives, it follows that there was no genuine bipolarity. Polish political scientist Roman Kuźniar, writing about the unforeseen collapse of the Soviet Union in 1989–1991, fittingly summarises the fiction of a global superpower clash as follows: 'This can be likened to a tree trunk full of dry rot, despite having a thin layer of bark, and which gives the appearance that all is well. A strong gust of wind is all it takes' (Kuźniar, 2005, p. 115).

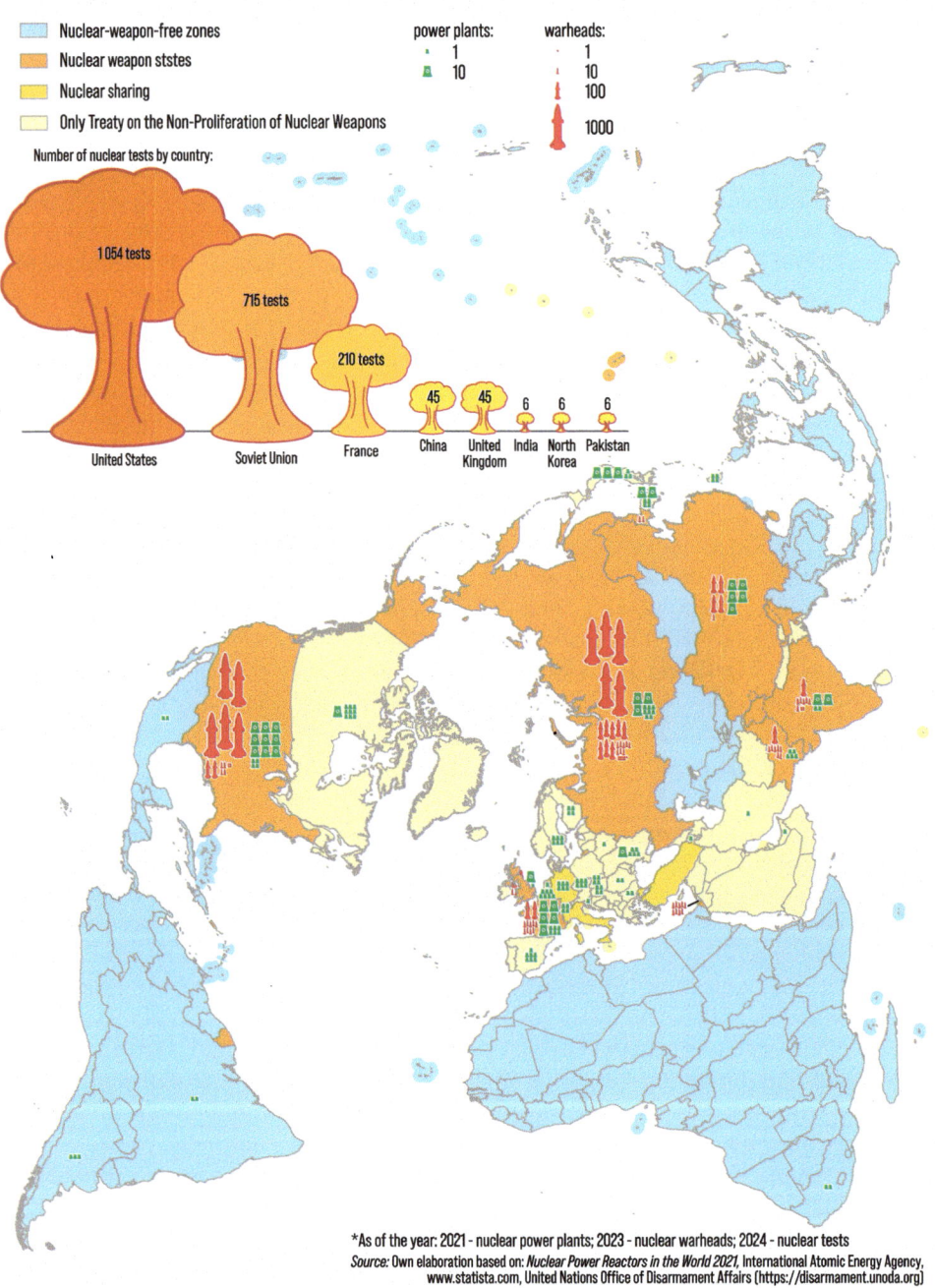

*Figure 8.19* Nuclear world: global Armageddon (warheads) vs clean energy sources (power plants)

Until June 1941, World War II was in fact a conflict between the international community, or the West (as broadly defined), and two criminal and revisionist totalitarianism regimes – one Nazi, the other communist (in this context, the plans of the English-French-Polish intervention on the side of Finland attacked by the USSR in 1939 should be recalled). Nazi Germany's invasion of communist Russia broke and blurred this clear dichotomy, as the totalitarian USSR became an ally of the democratic UK and United States: 'When the moment of truth arrived in 1941, allied leaders fighting for freedom and democracy did not hesitate to enlist one criminal in order to defeat another' (Davies, 1996, p. 899). One of the most glaring and shameful consequences of this was that Poland, the first casualty of WWII and the first country to fight Nazi Germany, was thrown to the wolves by the UK and the United States in clear violation of the 1941 Atlantic Charter in order to defeat the Third Reich and later had to accept the territorial reduction and Soviet vassalage that their allies had secretly agreed to with Stalin. The democratic West's approach to Poland was cold and cynical. It was an abandonment of common decency, a betrayal of inter-allied loyalty, and a repudiation of pacts willingly entered into. This conduct anticipated the formation of a new international order that would once more be built on an unequal and unjust peace and that would not, nor could ever be, the basis for lasting solutions. Only German totalitarianism was defeated and held responsible in 1945. In its origins and initial period (1939–1941), then, WWII was a war between the West and barbaric totalitarian regimes, and it only really ended in a partial victory for the West. While one totalitarianism was destroyed, the other sat at the victors' table. This elevated it to superpower status and made it a global partner and opponent of the West. The new peace was therefore not only unequal and unjust, as evidenced by the treatment of Poland and Czechoslovakia, but also defective. The Cold War, which was a clash between the West and Soviet totalitarianism, was therefore a de facto continuation of a conflict that had begun in 1939 and was resumed in 1945 after being put on hold in 1941 in the interest of their tactical alliance against the Third Reich. The incomplete victory of the non-totalitarian world in 1945 ushered in the third installment of the global conflict, known as the Cold War, de facto World War III. Again, in this view, the 30-year war (1914–1945) was actually a 77-year war (1914–1991), or an 'almost-hundred-year war' (Solarz, 2012, p. 64), which now in 2022–2024, may seem to have only apparently ended with the collapse of the Eastern Bloc and the Soviet Union in 1989–1991.

In the early 1990s it seemed that the cycle of the great 'almost-hundred-year' world war, in which the West had finally defeated two imperialist powers (German and Russian) and two totalitarianisms (Nazi and Soviet) in instalments, had finally come to a close. The period after 1989–1991, in which universal peace, democracy, and the pursuit of development and prosperity would supposedly reign under the watchful eye of the United States as 'a benevolent hegemon', appeared too quickly to many to have ended this period of mayhem into which the world had sunk in 1914. Shortly after the rivalry between the two blocs had ended, Norman Davies wrote: 'Only in the 1990s, with Germany reunited and the Soviet empire in a state of collapse, could the people of Europe resume the natural course of their development so rudely interrupted in the beautiful summer of 1914' (Davies, 1996, p. 900). Although it is worth pondering, when analysing historical processes in Europe up until 1914 and Japan before 1945 (as a consequence of the changes initiated by the 1868 Meiji Restoration, Japan became a quasi-European country), whether World War I and Japanese expansion in Asia and the Pacific were not the inevitable culmination of the 'natural course of their development' given the rapid pace of economic development and technical and organisational progress – which far outstripped the pace of moral progress – ushered in and shaped by the industrial revolution (see Solarz, 2012, p. 65). The best-known encapsulation of this wishful and naïve way of thinking comes from Francis Fukuyama, who wrote:

> We had reached the end of history. (. . .) And if we looked beyond liberal democracy and markets, there was nothing else towards which we could expect to evolve; (. . .) it was hard to find a viable alternative civilisation that people actually wanted to live in (. . .) there is only one system that will continue to dominate world politics, that of the liberal-democratic west. (. . .) time is on the side of modernity.
>
> (Fukuyama, 2001)

However, Fukuyama's conclusion that 'modernity is a very powerful freight train' which is difficult to derail (Fukuyama, 2001) may one day be proven correct, but it seems premature in 2022–2024. The bloody chaos in Yugoslavia, Somalia, and Rwanda in the early 1990s should suffice to undermine the expectation, let alone the hope, that – to paraphrase Robert Kagan – the world is slowly but inexorably 'entering into a post-historical paradise of peace and relative prosperity' (Kagan, 2003, p. 9).

At the turn of the 1990s, however, the course of events in the world seemed to confirm these predictions. The totalitarian communist systems of Central and Eastern Europe were dismantled in short order in 1989–1991, and the entire Eastern Bloc, along with the USSR, had collapsed by the end of 1991. In 1999–2007, all the post-communist countries of Europe (Bulgaria, Czechoslovakia [since 1993, Czechia and Slovakia], Poland, Romania, and Hungary), along with the former Soviet Republics of Lithuania, Latvia, and Estonia, joined the European Union (EU) and NATO (the GDR became part of the FRG in 1990 and was subsequently economically and militarily integrated with the West). In 1991, the United States, with a UN mandate, led an international coalition that drove Iraq out of Kuwait. There have also been successful attempts to shut down some Cold War 'fronts' since the late 1980s: German reunification and Namibian independence (1990) and the first multiracial elections, and the end of apartheid, in South Africa (1994).

Objectively speaking, the collapse of the USSR was wonderful news for the rest of the world – especially for those countries that had been colonised by the Russian Empire in either its 'white' or 'red' guise. The spectacular implosion of the Soviet empire at the turn of the 1990s had an economic as well as a territorial dimension. Russian gross domestic product (GDP) and per capita GDP declined precipitously in 1989–1998 (data.worldbank.org). The Soviet Union lost an enormous amount of territory when it lost the Cold War (name changes, along with words such as 'socialist', 'soviet', and 'federal', cannot conceal that what is denoted is the same old Russian Empire). The slow westward Russian floodtide from the 17th century temporarily ebbed in 1915–1939 before rapidly receding far to the east in 1989–1991. The Russian Empire was decolonised externally (its Central and Eastern European satellites threw off their subjection) and internally (the non-Russian Soviet Republics declared independence), albeit not completely, in 1989–1991. In 1991, Russia lost virtually all the Central Asian territories it had conquered in 1734–1895. The Caucasian territories of Armenia, Azerbaijan, and Georgia, which were won in 1801–1878 and had unsuccessfully attempted to break free of Russia after the 1917 revolution, likewise became independent when the USSR was dissolved. Leaving aside the temporary retraction of the borders of the Russian Empire in Europe in 1915–1939/40 as a consequence of WWI, which enabled several ethnic groups along the Baltic Sea–Black Sea axis to build or rebuild their own nations permanently or temporarily in 1918–1939, in 1990–1991 Russia lost a considerable portion of the territory it had gained during the Great Northern War (1700–1721), viz. Estonia and a large part of Latvia, as well as Crimea and a section of the northern Black Sea hinterland, annexed in 1783; several regions of Lithuania, Latvia, Belarus, and Ukraine, acquired when the Polish-Lithuanian Commonwealth was partitioned (1772–1793–1795); and those parts of Belarus and Ukraine taken from Poland and Lithuania in 1619–1667 (1686), together with the entire eastern portion of interbellum Poland, which it invaded and occupied in 1939 and which included territory that had been in the Austrian Partition in 1772–1918 and which had never been Russian. Russia also lost the following territories it had seized in 1940: the independent states of Lithuania, Latvia, and Estonia and the territories acquired by Romania in 1918, viz. Bessarabia, ceded to Russia by Turkey in 1812, and Northern Bukovina, ceded to Austria-Hungary by Turkey in 1775. In short, the Russian Empire receded to its 17th-century borders and 300 years of expansion were obliterated as a result of its failed rivalry with the United States. When Russia's Cold War defeat, which cost it 24% of its territory, is compared with the WWI defeat of Hungary (71%) and Germany (13%), the emergence of a revisionist regime in Russia in response to new borders and a new international order comes as no surprise. From a Russian imperial standpoint, Vladimir Putin was therefore correct when he stated in April 2005 that the breakup of the USSR was the greatest geopolitical catastrophe of the 20th century (Krajewski, 2022).

*Figure 8.20* Development of Russia's western border: 1618–2024

*Figure 8.21* Russia – borders and territory: 1914–2024

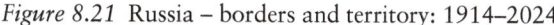

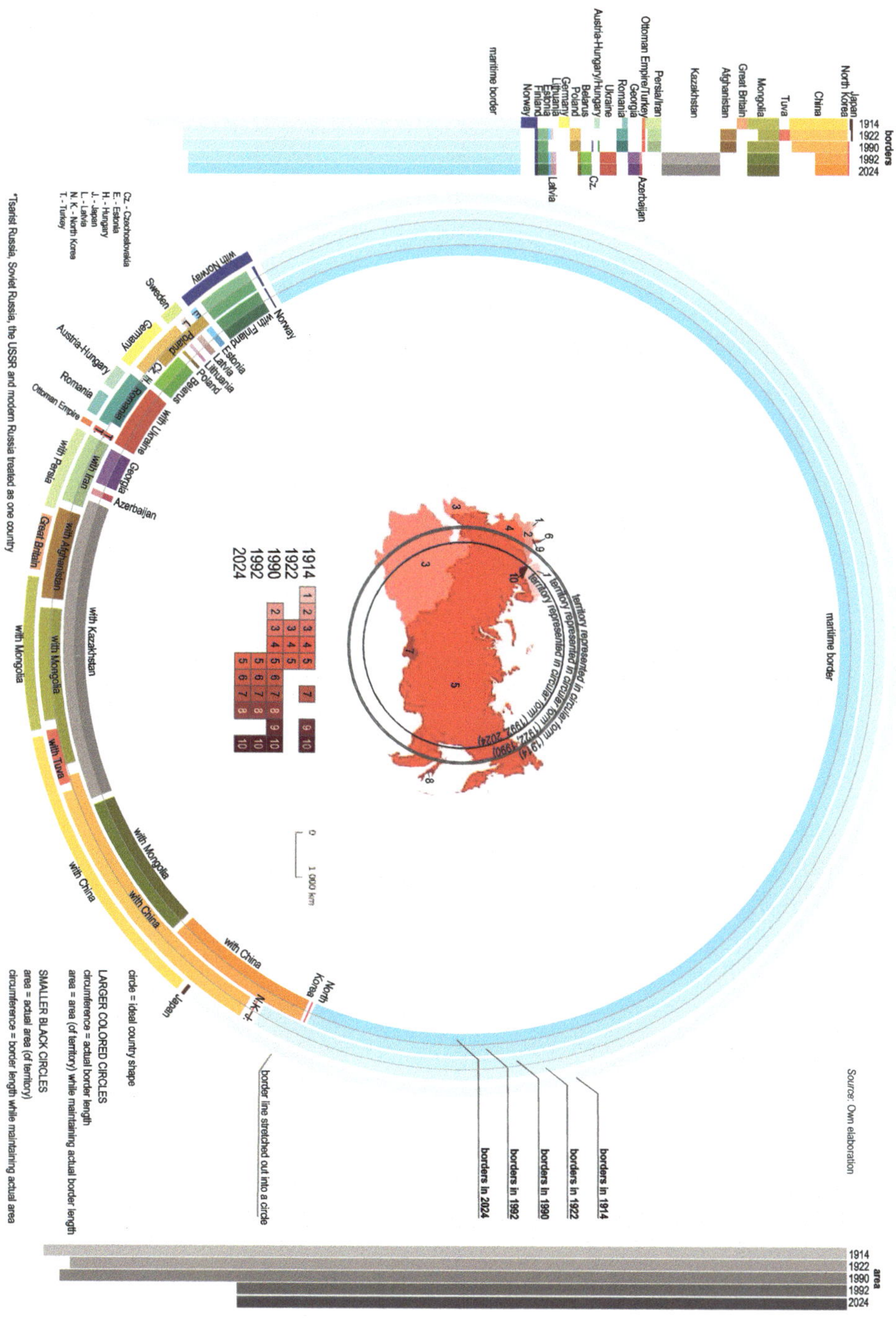


Source: Own elaboration

The undeniably positive development in international policy resulting from the end of the Cold War, however, was neither universally successful nor beneficial. Nor was it universally accepted and supported. On the other hand, the 1990s witnessed failure of the UN mission under the leadership of the United States to stabilise Somalia (1992–1995), brutal wars following the breakup of Yugoslavia (1991–1995), a genocide in Rwanda (1994), and the first Taliban government in Afghanistan (from 1996).

**Yugoslavia**

— on the eve of its disintegration (1990)
— after the peace agreement in Dayton, OH, USA in 1995
— after the separation of Kosovo in 1999
— after Montenegro's secession in 2006
— borders of other countries in the former Yugoslavia in 1995

**Bosnia and Herzegovina after the peace agreement in Dayton, USA in 1995**

Federation of Bosnia and Herzegovina

Republika Srpska

Brčko District

\* In 2003, the Federal Republic of Yugoslavia became Serbia and Montenegro. In 2006, Montenegro declared independence.

*Source*: own elaboration

*Figure 8.22* Evolution of borders in Yugoslavia: 1918–2024

With the end of the Cold War the United States attained, and was to maintain, such predominance over the rest of the international community that the world had seemingly entered an epoch of *pax americana*, where it could enjoy relative peace and focus on development and building prosperity. Any internal disturbances in the international system would be peripheral and amenable to diffusion with some prompting, or as a last resort, force, on the part of the United States. Although the United States came to dominate the international system in 1989–1991, however, this dominance was never absolute and was challenged over time. The opening assault on the post–Cold War status of the United States comprised the four coordinated terrorist attacks in New York and Washington on 11 September 2001. The 9/11 attacks were met with US-led coalitions invading and occupying Afghanistan (2001–2021) and Iraq (2003–2011). The United States asserted its military dominance in both instances, but proved incapable of eliminating the threat of radical Islam. Nor was it able to engraft its political and developmental model onto either country. Radical Islam thus became Washington's first major antagonist, and this rebellion was directed against the US domination that resulted from the Cold War. The second, and so far, the most serious, challenge to US hegemony has come from the People's Republic of China (PRC). The PRC has not only filled the power vacuum left by the defeat of the USSR in the 1945–1991 global competition but has been rapidly modernising and has recorded prodigious economic growth for the past 45 years. The next to challenge US post–Cold War dominance, although it did not do so openly until 2022, was Russia. Its behaviour results both from its disagreement with the global political and territorial decisions of 1989–1991 and from the fact that its defeat was not complete and unambiguous like that of Germany in 1945. Nevertheless, the only real threat and challenge that Russia – essentially an oil pump with an atom bomb (Solarz, 2009, pp. 132–134, 2014b, pp. 144–145) – still poses to the international community lies in its nuclear arsenal. This former superpower no longer poses a serious threat, either militarily (as evidenced by the ongoing Russo-Ukrainian War) or economically (data.worldbank.org).

Vladimir Putin burst onto the Russian political scene in the late 1990s. Appointed prime minister in 1999 (and destined to become president the following year), Putin immediately set about reversing the consequences of Russia's Cold War defeat. From the perspective of a quarter of a century, his most profound, long-term aim seems always to have been to 'make Russia great again' by returning to imperial policy and reinstating the country's international position, restoring its influence, and regaining its lost territories. After 1991, however, Russia's political elites proved incapable of critically reflecting on Russian history and wallowed in the geopolitical trauma brought on by the events of 1989–1991 (the collapse of the external empire and the USSR; the retreat of the empire to the borders of the Russian Union Republic of the Soviet Union; the loss of superpower status). Putin intended to pursue his objective primarily through violence and conquest, eschewing any deeper reconstruction of Russia, as is borne out by his 25 years as either president or prime minister. The political borderland between East and West in Central and Eastern Europe, on the eve of the initial Russian aggression towards Ukraine in 2014, could be categorised as (to use US election terminology) 'safe states' (for the West, these were definitely Poland, Romania, Lithuania, Latvia, and Estonia; for Russia, Belarus was a surety) and 'swing states' or 'battleground states' (especially Ukraine, excluding Russia, the second largest and sixth most populous country in Europe). Neither the West nor Russia could have predicted the final outcome of their marathon chess match in Ukraine and other 'battleground states'. Ukrainian politics and society had been sharply divided between pro-Russian and pro-Western factions since independence in 1991. Ukraine made three radical turnarounds in 2004–2014: the pro–West Orange Revolution of 2004, the election of a pro-Russian president in 2010, and his removal from office in 2014 following the Euromaidan Uprising (November 2013 to February 2014). This time, however, turning westward provoked Russian intervention. The Russian invasion of Georgia (2008) was the first clear warning of the impending historical storm, but the next was its annexation of Crimea (2014) and the establishment of two separatist puppet republics of Donetsk and Luhansk in the east of Ukraine in 2014, formally annexed by Russia in September 2022.

Putin's step-by-step attempt to restore the Russian Empire by expanding into the former Soviet sphere has been accompanied by actions intended to neutralise Western resistance. The two Nord Stream gas pipelines (NS1, NS2), which run under the Baltic to directly connect Russia and Germany, were meant to play a special role in this plan. Construction of the pipelines, jointly supported by Russia and several West European countries, began in 2005 and was completed in 2011 (NS1) and 2021 (NS2), although the pipelines are not operational as a consequence of the Russo-Ukrainian War and of damage sustained in 2022. The economic, environmental, and security interests of other countries along the Baltic, as well as those in the buffer zone between Germany and Russia, were disregarded and their protests ignored. Securing natural gas supplies was supposed to make Western Europe largely indifferent to Russian expansion. Without the pipelines, distribution to Western Europe could have been interrupted as a result of Russia blocking land supply routes to politically or economically pressure its neighbours east of Germany or as a consequence of military operations in CEE. This, in turn, could have forced the West to act to defend its interests and its allies in the eastern part of the continent.

In 2022, the conflict in Ukraine escalated into a full-scale 'hot war' that constitutes a full and open challenge to the world peace and international order that resulted from the USSR's Cold War defeat. Russia's invasion of Ukraine was a direct and foreseeable consequence of its discontent with the new international order. This includes the way it appeared on the world political map following its failed confrontation with the United States. In 2021, the Russian dependency Belarus had (unsuccessfully) attempted to spark a migration crisis on the Polish, Lithuanian, and Latvian borders by inciting and assisting illegal immigrants to cross an external NATO and EU border, thereby triggering a hybrid war that was ongoing at the time of publication (2024). Belarus was initially thought to be retaliating for the sanctions imposed by the EU in response to its rigged 2020 presidential election, but after 24 February 2022, it became apparent that it probably had another, more sinister motive. Belarus was attempting to destabilise the EU and NATO internally in readiness for Russia's impending invasion of Ukraine. Had Belarus been successful in its attempt to have migrants spilling over from Central and Eastern Europe to the westernmost reaches of the continent, it would probably have stymied any Western response to Russian aggression.

The Western refusal to make any concessions to the de facto ultimatum that Russia delivered to the United States and NATO at the end of 2021 (Menkiszak, 2021) and which would have resulted in a 'fundamental overhaul of the European security order in favour of Russia' (Menkiszak, 2021), effectively reversing the history of the previous 30 years, was in all likelihood merely a pretext for Russia to launch a full-scale land, sea, and air invasion of Ukraine without a declaration of war on the morning of 24 February 2022 (reminiscent of Nazi Germany's 1939 invasion of Poland). As was

WWII, Putin's policy, including his aggression towards Ukraine, is a natural outgrowth of refusing to accept the verdict of history. The Russo-Ukrainian War now seems to have extended the 20th-century time of troubles – a 77-year conflict (1914–1991) in a 100-year world war (1914–2024 and beyond?) – after a relatively peaceful intermission at the turn of the millennium, although it is still not known whether this is merely the latest paroxysm in the phase of history that began in 1914 or whether, given NATO support for Ukraine, it is the beginning of a fourth global clash (i.e., WWIV) whose length, character, dynamics, and outcomes (including those reflected on the world political map) are not known. Regardless of how this question will be answered, Russia has been a permanently toxic member of the international community for centuries. Russian imperialism has either had a hand in or been directly responsible for every tragic historical turning point over the past 100 years – Sarajevo, the Molotov-Ribbentrop Pact, the iron curtain, the aggression towards Ukraine. This assessment is not meant to absolve Germany of its terrible liability – 'Europe's newest, most dynamic, and most disgruntled nation state' (Davies, 1996, p. 900). The heart of the troubles, however, was not Germany alone, but Germany in tandem with Russia – 'from the start, the major duel over Europe's future lay between Germany and Russia' (Davies, 1996, p. 900).

The destructive role of Russia in the international (European) system was pointed out as early as 1878 by the Polish geographer Wincenty Pol, who foresaw not only the iron curtain in general, but he also roughly placed it in the correct geographical location nearly 70 years before its creation. Pol maintained that unless Europe grasped Russia's detachment from European civilization, it would see the return of a very real and strong threat of military invasion from the East. 'Unless it is understood that Europe ends at the Dnieper and Poland, the Tsar will mark out a new map of Europe along the Adriatic coast and on the borders of the western Slavic region with a whip' (Pol, 1878, p. 22) (sic) (the West Slavic ethnicities are the Poles, Czechs, and Slovaks, all of whom have their own countries, and the Lusatians, who live in east Germany close to the border with Poland and the Czech Republic, in the former GDR). Russia's transformation into a 'quite ordinary country' might be an opportunity to put an end to the time of troubles, i.e., the cycle of hot and cold world wars that began in 1914 and in which Russia has always played a leading role. However, it is not easy to be optimistic. In contradistinction to Germany in 1945, Russia has a nuclear arsenal that precludes any externally imposed change, and change from within seems unlikely at the moment. History also shows that it is only a matter of time before a strong country intimidates, blackmails, or attacks its next victim, invariably a weaker country, and a great imperialist power has no principles in this regard and can therefore be counted on to do so. Most people obviously want to live in peace, but it is worth recalling a speech that Polish foreign minister Józef Beck gave on 5 May 1939, when the pressure from the Third Reich that was to find relief in the invasion of 1 September was increasing: 'Peace is a precious and a desirable thing. [. . .] But peace [. . .] has its price, a high but a measurable one. We in Poland do not know the concept of peace at any price' (Kuźniar, 2009, p. 12). So, what is the price of peace in Ukraine now? The international community's recognition of the illegal annexation of Crimea and eastern Ukraine? How about the whole country? Or 'only' the loss of its right to political self-determination and independent development? A price paid by others is still a price. The history of the period 1938–1940 clearly shows the futility of the UK and France's attempts to palm the costs of peace off onto Czechoslovakia and Poland. It brought them no benefits. They failed to halt the expansion of Nazi Germany, they failed to preserve the peace, and they lost their great power status (and France lost the 1940 war in the bargain). By contrast, in 1945 the USSR, having already been rewarded for its joint responsibility in starting WWII (by concluding, and executing, the Molotov-Ribbentrop Pact, including its invasion of Poland in 1939), received more than it had sought to gain from its pact with the Third Reich from the Western powers. This reward in no way signalled – as was the case with the Third Reich – a renunciation of its policies of confrontation, aggression, unconstrained territorially, and aimless imperialism (Ferro, 1997, p. 24), which led to the Cold War. Similarly, rewarding this notorious aggressor from the borderland of Europe and Asia yet again in 2024 or later will not bring, as British Prime Minister Neville Chamberlain put it in 1938, 'peace in our time'. If history is truly the master of life and Russia's past behaviour is any guide, then it is more likely to result in an escalation of demands and further aggression. To quote once more from Beck's speech of May 1939: 'Two conditions are necessary for this word [peace – M.W.S.] to be of real value: (1) peaceful intentions, (2) peaceful methods of procedure' (Dziurok *et al.*, 2010, p. 90). Unfortunately, whether contemporary Russia has peaceful intentions and whether it uses peaceful means of procedure towards its neighbours are purely rhetorical questions at this stage.

While these consecutive worldwide conflicts were not all driven by identical mechanisms, many of their causal factors were duplicated in different proportions to create new explosive mixtures. 'In history – let's say it again – certain situational arrangements and patterns' are repeated, even if 'in most cases [actually, always – M.W.S.] history is the catalogue of the premieres themselves. [. . .] "the same" is never "the same" and new versions of almost identical situations can produce different results' (Mieroszewski, 1974, p. 3). Simply put, the history of humankind is marked by the paradox of the coexistence of continuity and change. For example, the United States has played a major role in every scene of this Long War (1914–1991):

Faced by German expansionism, and then by the twin hydras of communism and fascism, the democratic Western Powers could only survive by calling in the USA – first in 1917–1918 and then in 1941–1945. After 1945 they relied very largely on American muscle to withstand the challenge of a bloated Soviet empire.
(Davies, 1996, p. 900)

The situation has reasserted itself for the fourth time in 100 years since 24 February 2022. Once more, 'certain situational arrangements and patterns' are being repeated (Mieroszewski, 1974, p. 3). The United States has become the guarantor of the security of the European part of NATO, especially its eastern flank, in as clear a manner as it had been prior to 1989/1991, and Ukraine could not continue fighting without its assistance.

It should also be noted that the period 1914–2024 was a causal chain marked by continuity and change so far as decolonisation is concerned. First, WWI left the European colonial powers debilitated and marked the beginning of the loss of their power and political leverage in international relations. This triggered the second wave of decolonialisation (North and Latin America being the first) and led to the world political map being redrawn: Russia and Germany had lost all or part of their colonies. Second, following on from WWI (the loss of the colonial territories was not accepted by them either), WWII sparked the third wave of decolonisation, which mainly affected the UK, France, the Netherlands, Belgium, Italy, and Portugal. This was a powerful weapon for the USSR in its Cold War confrontation with the US-led West. Finally, the Cold War defeat of

Russia caused the fourth wave of decolonisation at its expense. Russia's refusal to accept its results culminated in its invasion of Ukraine in 2022.

Last but not least, even if Russia took the first step in an open and overtly provocative way in 2022 to reverse the US victory in the Cold War, the PRC will undoubtedly play a key role in the latest instalment of the new Hundred Years' War. Until the end of the 1970s, the PRC was just one of many Third World countries. The turn of the 1980s was a watershed in its history. The country began to develop rapidly as a result of Deng Xiaoping's economic reforms, and its prodigious growth since 1978 has made it a major world exporter. China's precipitate development, spanning almost half a century, has not only allowed it to outclass Russia but also to effectively pursue the United States. The PRC has become a serious rival to the United States and a challenger to its global power, and consequently, there is increased friction between Washington and Beijing. Significantly, the simultaneous failure and breakup of the USSR broke the bond of Sino-American cooperation. The rapid development of the Chinese economy together with the collapse of Soviet and Russian power mean that the PRC no longer has to feel threatened by Russia. Various 'situational arrangements and patterns' were therefore formed in the US-PRC-Russia triangle after 1945: the PRC was a junior partner in an anti-US alliance with the USSR from 1949, it was in conflict with both the United States and the USSR in the 1960s, and it was working with the United States to the detriment of the USSR in the 1970s and 1980s. The PRC might now be renewing its anti-US alliance with Russia, only this time as the senior partner.

# 9 The borders

This section contains several maps showing the evolution of the borders of CEE in 1918–2024. These can be thought of as 'snapshots' that capture borderlines at specific moments in history. The chapter is supplemented by a series of five maps showing the balance of territorial transformations in Europe in relation to WWI, the interwar period, WWII, the Cold War, and the period that came after it. In order to highlight these territorial changes, the latter maps only show the territorial changes with no regard to the entire politico-geographical map of Europe.

DOI: 10.4324/9781003406440-9

*Figure 9.1* Borders in CE Europe: April 1918

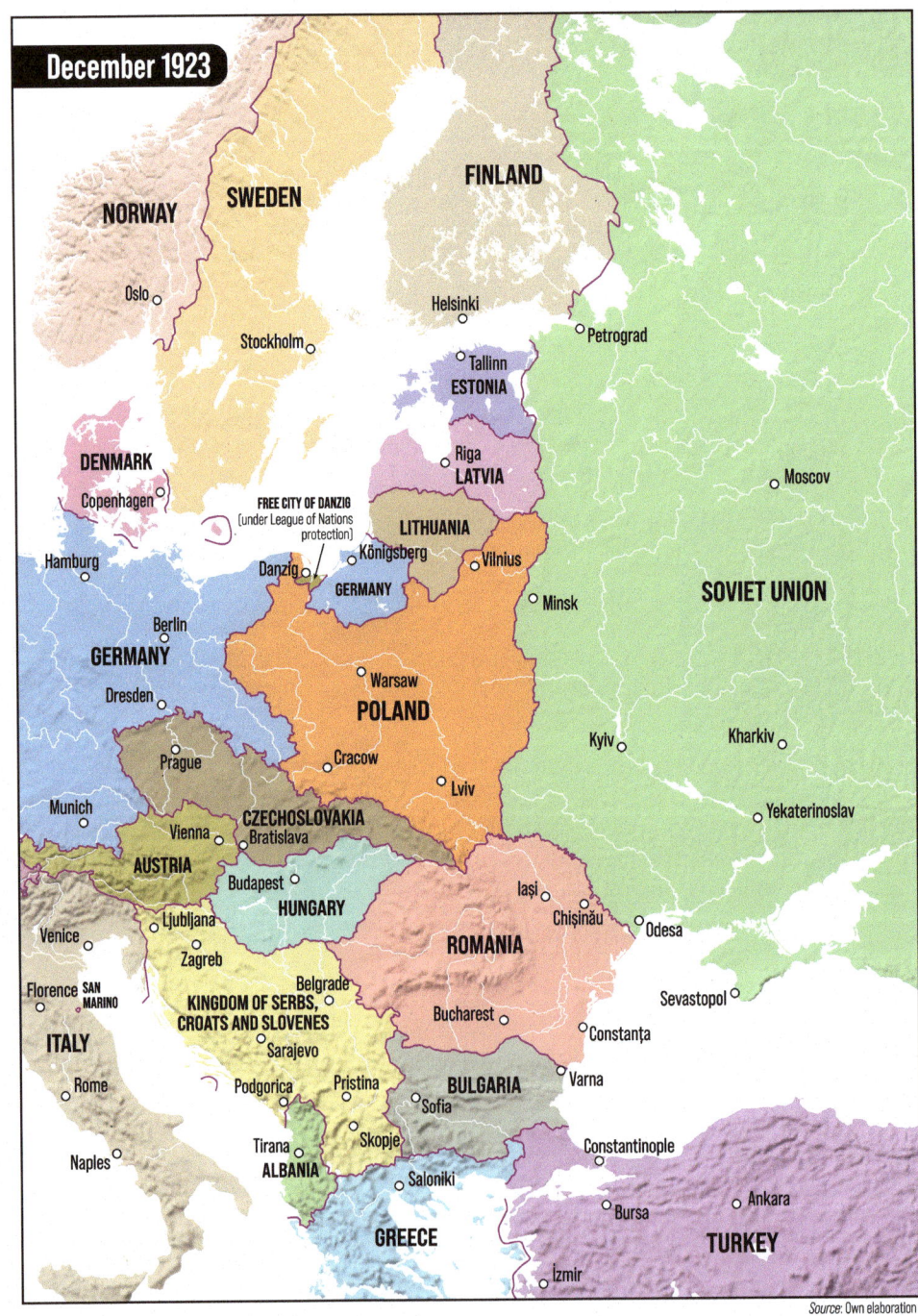

*Figure 9.2* Borders in CE Europe: December 1923

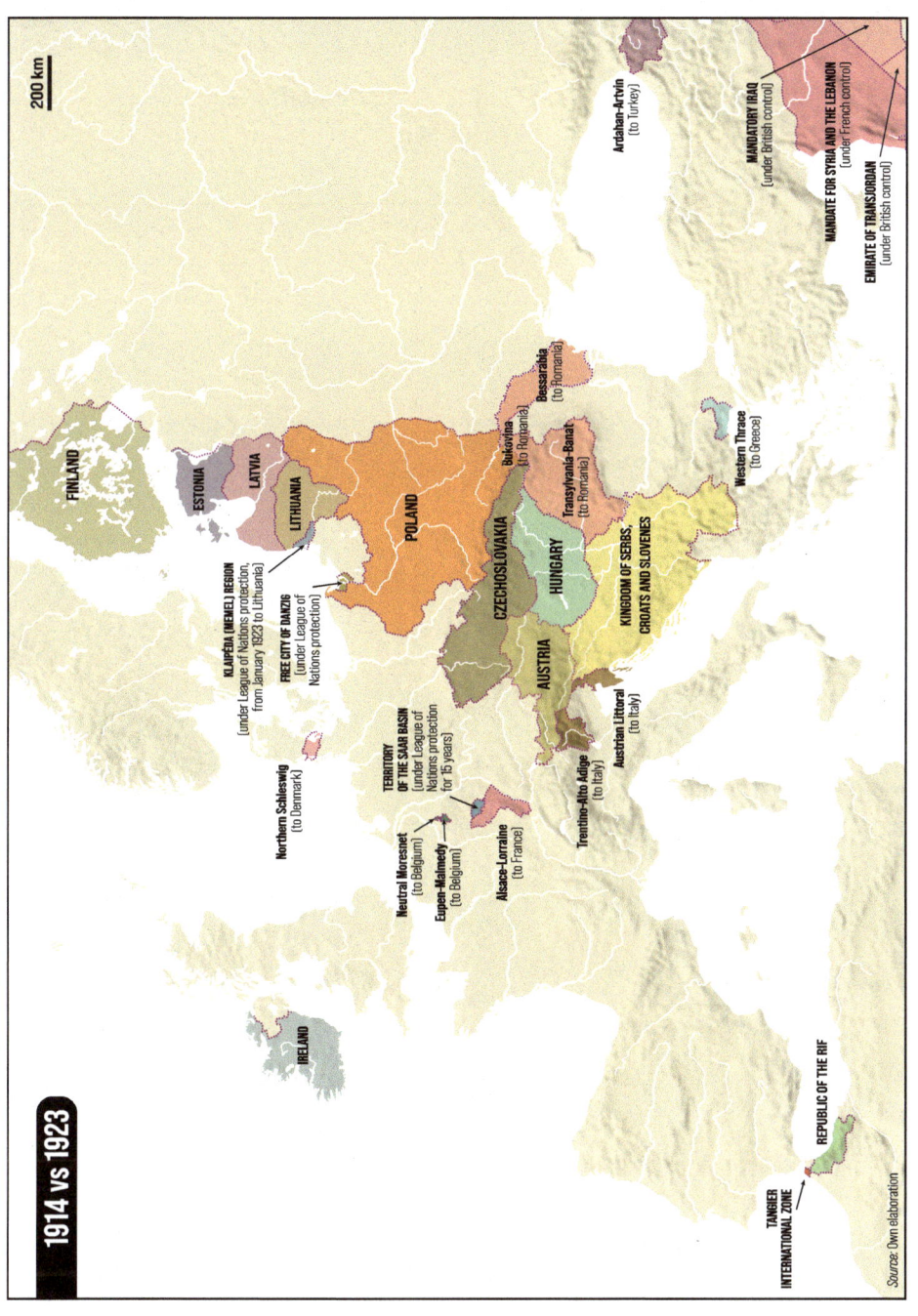

*Figure 9.3* Territorial changes in Europe: 1914 vs. 1923

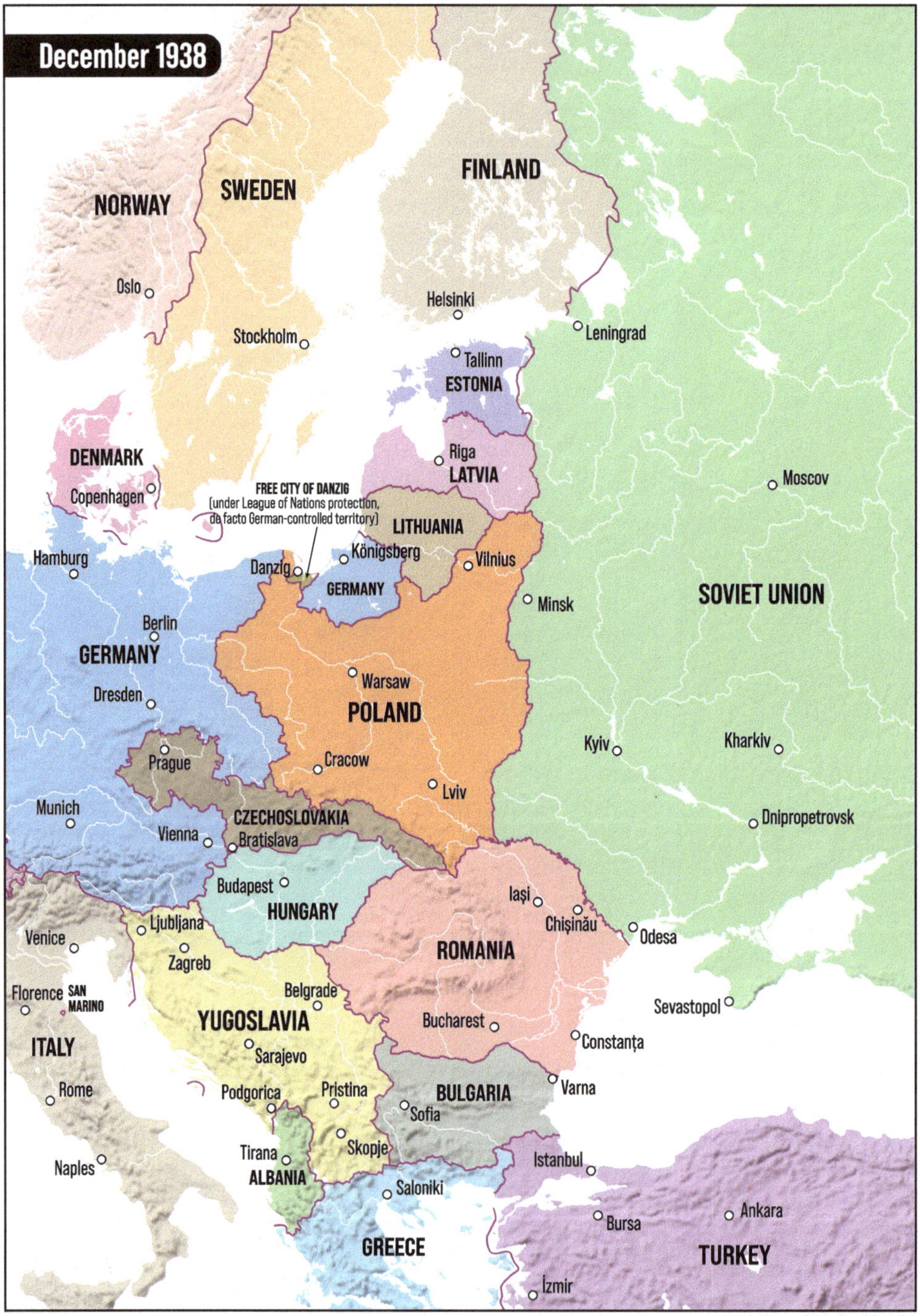
December 1938
NORWAY
SWEDEN
FINLAND
Oslo
Stockholm
Helsinki
Leningrad
Tallinn
ESTONIA
DENMARK
Copenhagen
Riga
LATVIA
Moscov
FREE CITY OF DANZIG
(under League of Nations protection,
de facto German-controlled territory)
LITHUANIA
Hamburg
Danzig
Königsberg
Vilnius
GERMANY
Minsk
SOVIET UNION
Berlin
GERMANY
Warsaw
Dresden
POLAND
Kyiv
Kharkiv
Prague
Cracow
Lviv
Munich
CZECHOSLOVAKIA
Dnipropetrovsk
Vienna
Bratislava
Budapest
Iași
HUNGARY
Chișinău
Ljubljana
Odesa
Venice
ROMANIA
Zagreb
Florence
SAN MARINO
Belgrade
Sevastopol
YUGOSLAVIA
Bucharest
ITALY
Sarajevo
Constanța
Rome
Podgorica
Pristina
BULGARIA
Varna
Sofia
Tirana
Skopje
Istanbul
Naples
ALBANIA
Saloniki
Bursa
Ankara
GREECE
TURKEY
İzmir

*Source*: Own elaboration

*Figure 9.4* Borders in CE Europe: December 1938

*Figure 9.5* Borders in CE Europe: August 1939

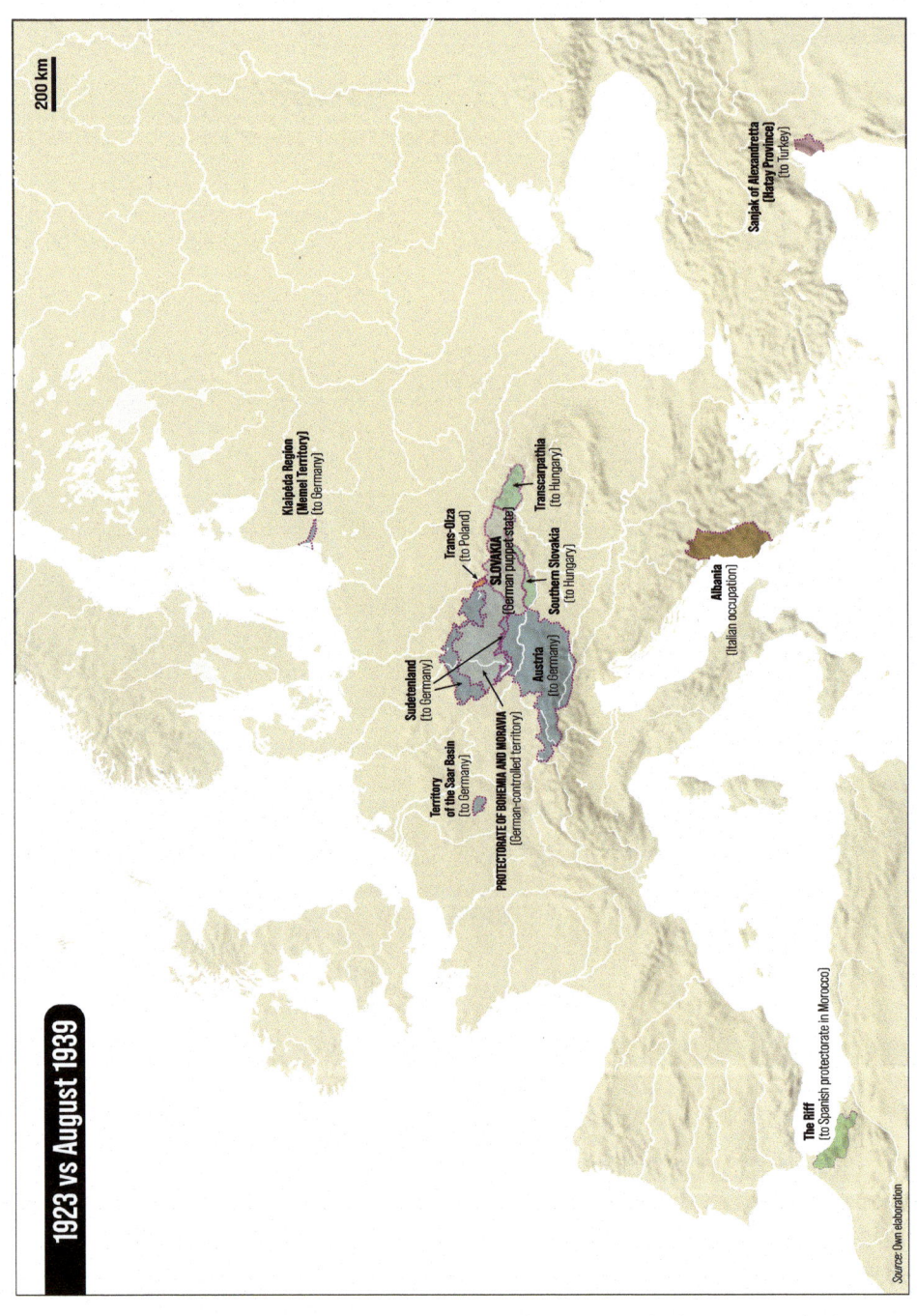

*Figure 9.6* Territorial changes in Europe: 1923 vs August 1939

*Figure 9.7* Borders in CE Europe: October 1939

*Figure 9.8* Borders in CE Europe: September 1940

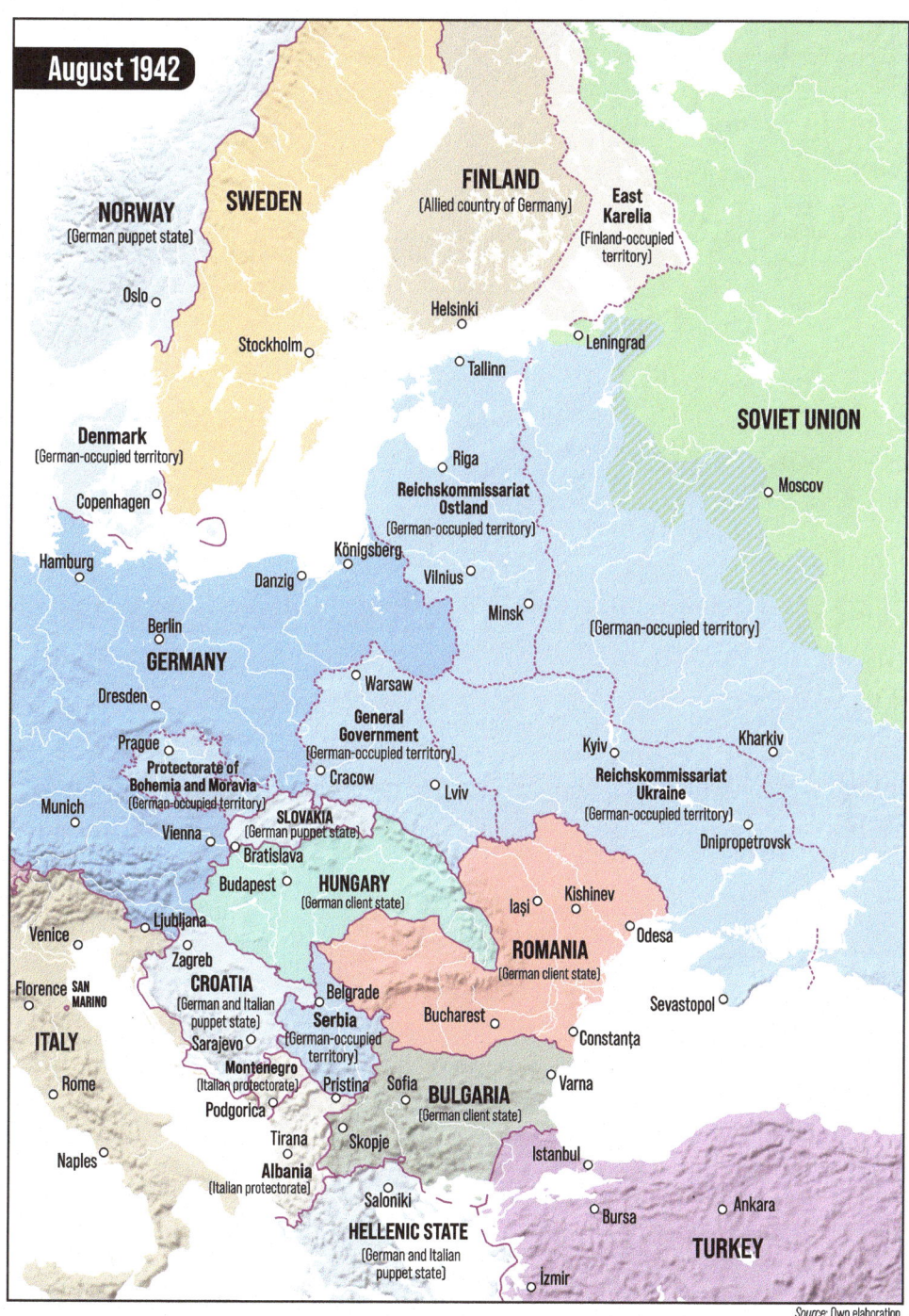

*Figure 9.9* Borders in CE Europe: August 1942

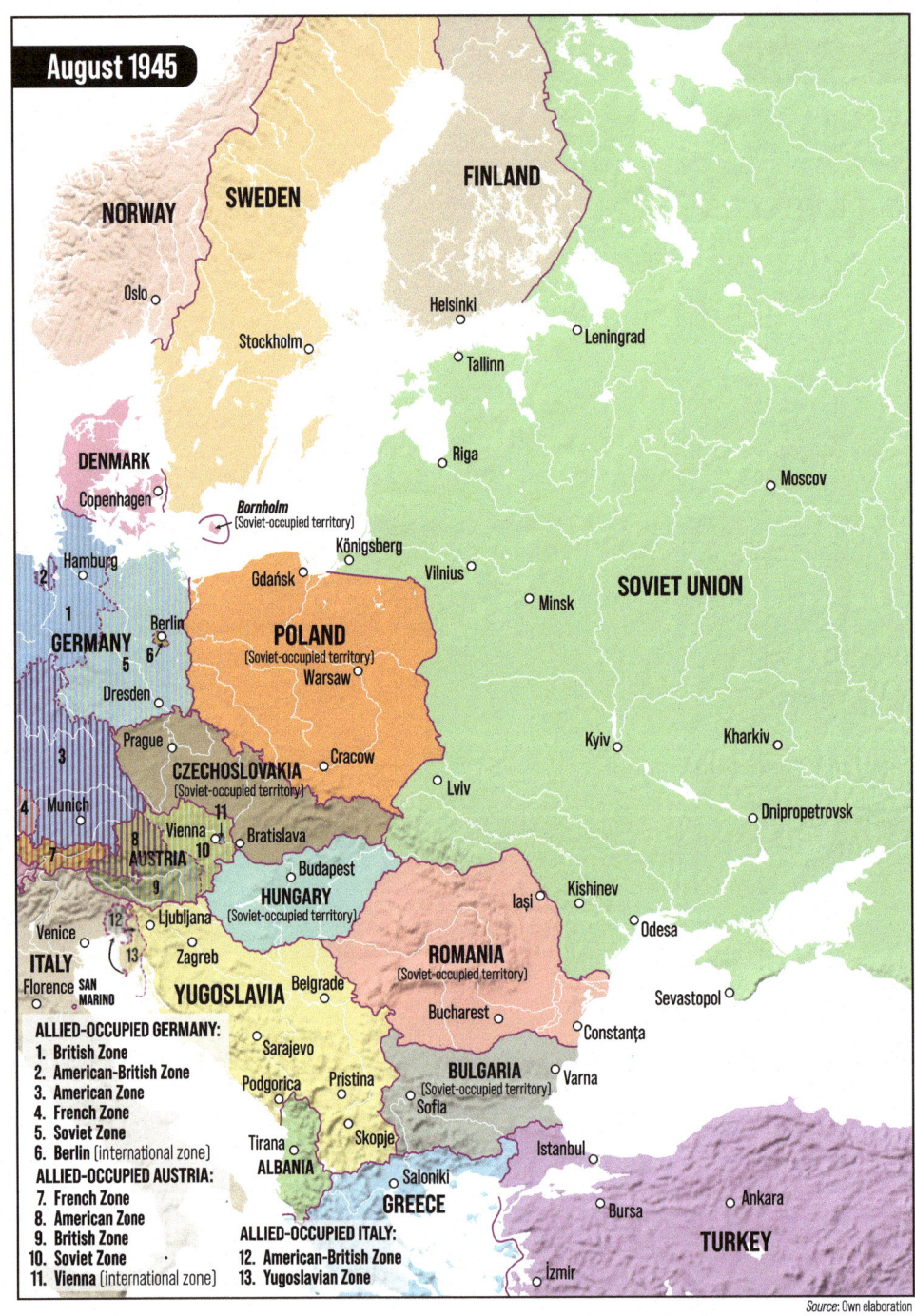

**August 1945**

NORWAY
Oslo

SWEDEN
Stockholm

FINLAND
Helsinki

Leningrad

Tallinn

DENMARK
Copenhagen

Riga

Moscov

*Bornholm*
(Soviet-occupied territory)

Königsberg

Hamburg

Gdańsk

Vilnius

SOVIET UNION

Berlin
GERMANY

POLAND
(Soviet-occupied territory)
Warsaw

Minsk

Dresden

Prague

Cracow

Kyiv

Kharkiv

CZECHOSLOVAKIA
(Soviet-occupied territory)

Lviv

Munich

Dnipropetrovsk

Vienna
AUSTRIA
Bratislava

Budapest

HUNGARY
(Soviet-occupied territory)

Kishinev

Venice

Ljubljana

Odesa

ITALY
Florence SAN MARINO

Zagreb

Belgrade

ROMANIA
(Soviet-occupied territory)

Iaşi

Sevastopol

YUGOSLAVIA

Bucharest

Constanţa

**ALLIED-OCCUPIED GERMANY:**
1. **British Zone**
2. **American-British Zone**
3. **American Zone**
4. **French Zone**
5. **Soviet Zone**
6. **Berlin** (international zone)
**ALLIED-OCCUPIED AUSTRIA:**
7. **French Zone**
8. **American Zone**
9. **British Zone**
10. **Soviet Zone**
11. **Vienna** (international zone)

Sarajevo

Podgorica

Pristina

BULGARIA
(Soviet-occupied territory)

Varna

Sofia

Tirana

Skopje

Istanbul

ALBANIA

Saloniki

GREECE

**ALLIED-OCCUPIED ITALY:**
12. **American-British Zone**
13. **Yugoslavian Zone**

Bursa

Ankara

TURKEY

İzmir

*Source*: Own elaboration

*Figure 9.10* Borders in CE Europe: August 1945

*Figure 9.11* Borders in CE Europe: 1949

*Figure 9.12* Borders in CE Europe: January 1955

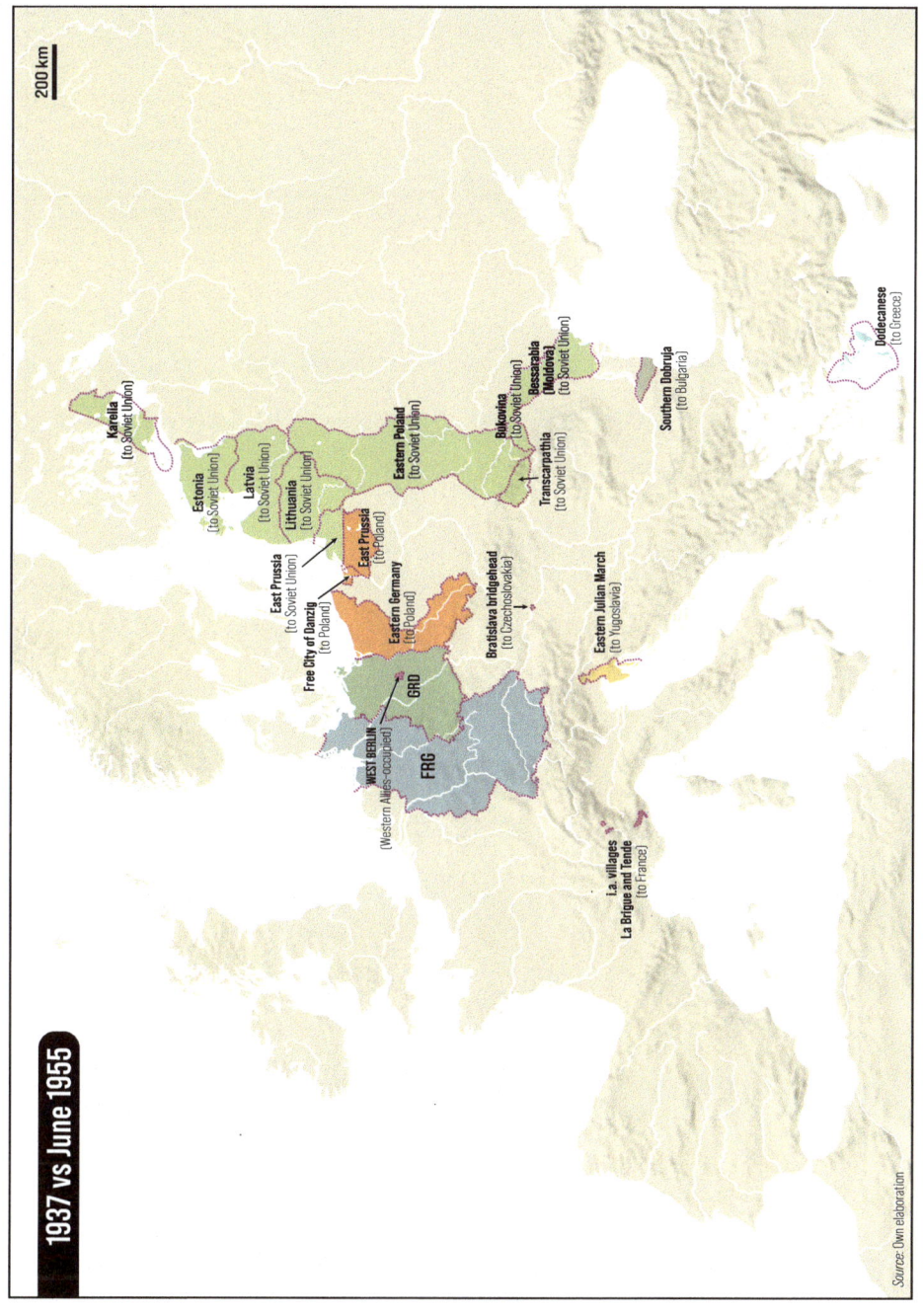

*Figure 9.13* Territorial changes in Europe: 1937 vs June 1955

*Figure 9.14* Borders in CE Europe: November 1990

*Figure 9.15* Borders in CE Europe: January 1992

*Figure 9.16* Borders in CE Europe: January 1993

*Figure 9.17* Borders in CE Europe: January 1996

*Figure 9.18* Territorial changes in Europe: 1989 vs June 1999

*Figure 9.19* Borders in CE Europe: December 2013

*Figure 9.20* Borders in CE Europe: March 2015

Figure 9.21 Borders in CE Europe: 30 March 2022

*Figure 9.22* Borders in CE Europe: 1 January 2024

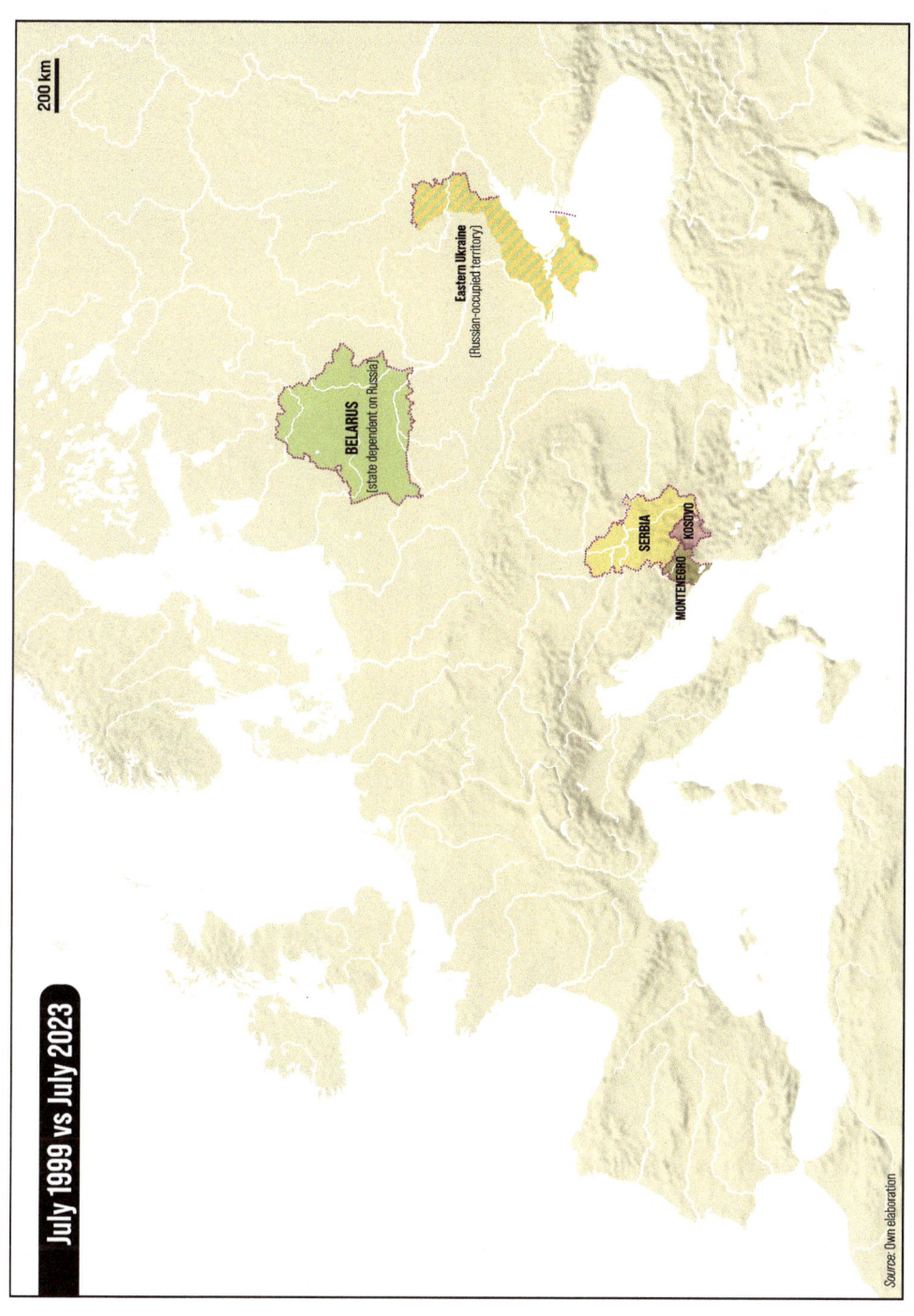

*Figure 9.23* Territorial changes in Europe: July 1999 vs July 2023

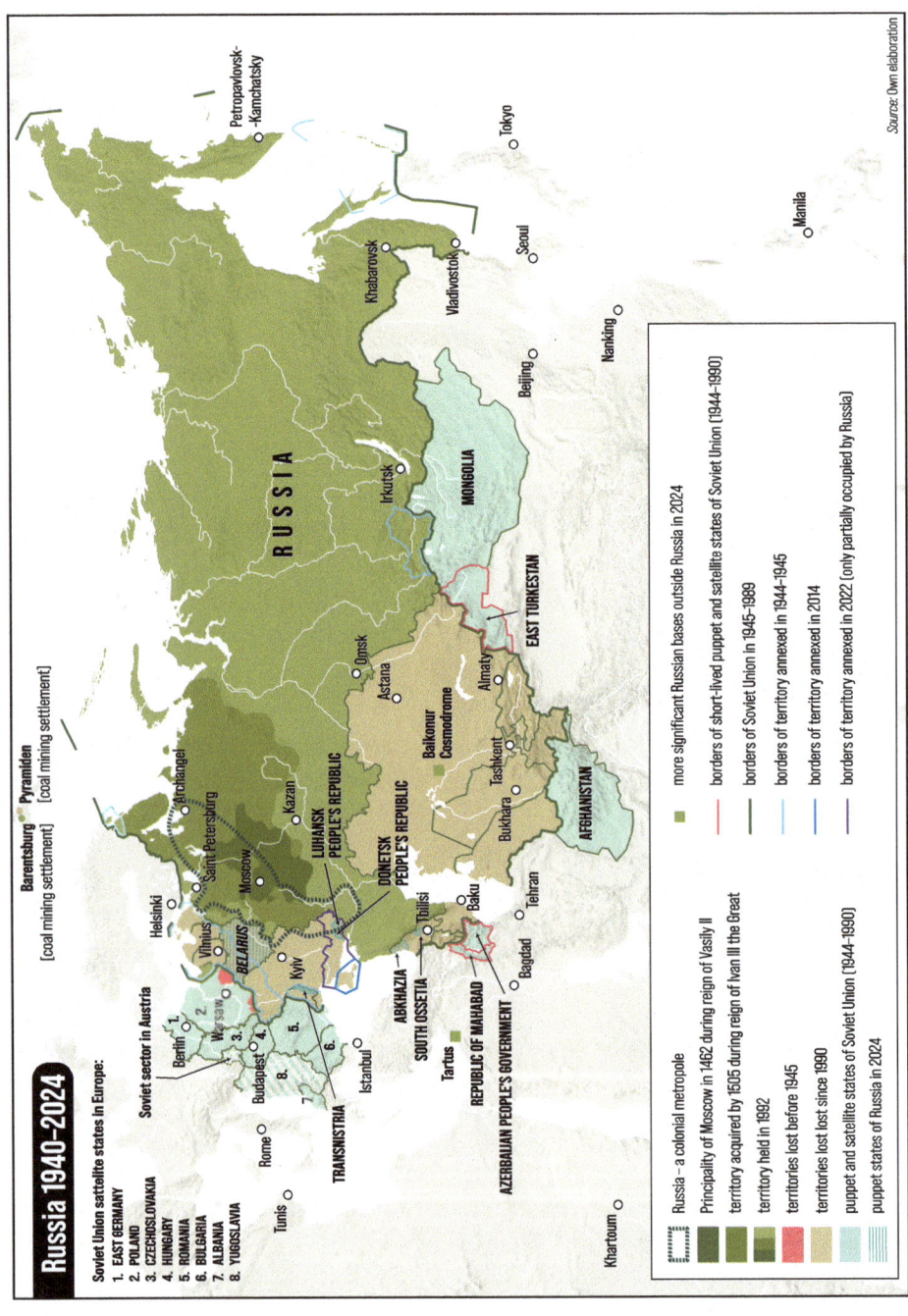

*Figure 9.24* Red and White Russia: 1940–2024

# 10 The power systems and arrangements

This chapter comprises a group of cartographic studies depicting former and current political blocs in international relations or different formations of countries (including e.g., West vs East, North vs South), including the history of selected international organisations (e.g., NATO, the EU, the Warsaw Pact) and balance maps drawn for the period after 1945. In addition to raising awareness of international connections and explaining the behaviour and interactions observed in the international community, the former maps are meant to serve as 'navigation maps' that facilitate the interpretation of the other maps in the atlas.

The latter collate selected mutually hostile alliances, viz. the US/NATO vs USSR/Russia/Warsaw Pact (also taking into account China), including Ukraine with regard to recent history, in terms of economic (GDP and per capita GDP), population, and military (total military expenditure, military expenditure as a percentage of GDP, and military expenditure per capita) data. These also enable the potentials within the blocs, as well as those between individual countries, to be compared. Again, only a selection of countries is shown. This is done so as to not lose sight of what is most important amid the sheer volume of information and the noise of current history. Unfortunately, not all the maps contain complete data and comparisons. After all, maps are merely prisoners of the data on which they are based and without which they cannot be made.

Balance maps cover the following seven periods:

1950–1956: from the outbreak of the Korean War to the Soviet intervention in Hungary;
1957–1980: from the Soviet intervention in Hungary to the Soviet intervention in Afghanistan (1979) and the birth of the 'Solidarity' movement in Poland (1980);
1981–1990: from the suppression of 'Solidarity' to the end of the Cold War;
1991–1998: from the end of the Cold War to the first eastward expansion of NATO (1999);
1999–2013: from the first eastward expansion of NATO to the outbreak of the Russo-Ukrainian war (2014);
2014–2021: from the outbreak of the Russo-Ukrainian war to full-scale Russo-Ukrainian confrontation (2022);
2022: outbreak of the full-scale Russo-Ukrainian war.

DOI: 10.4324/9781003406440-10

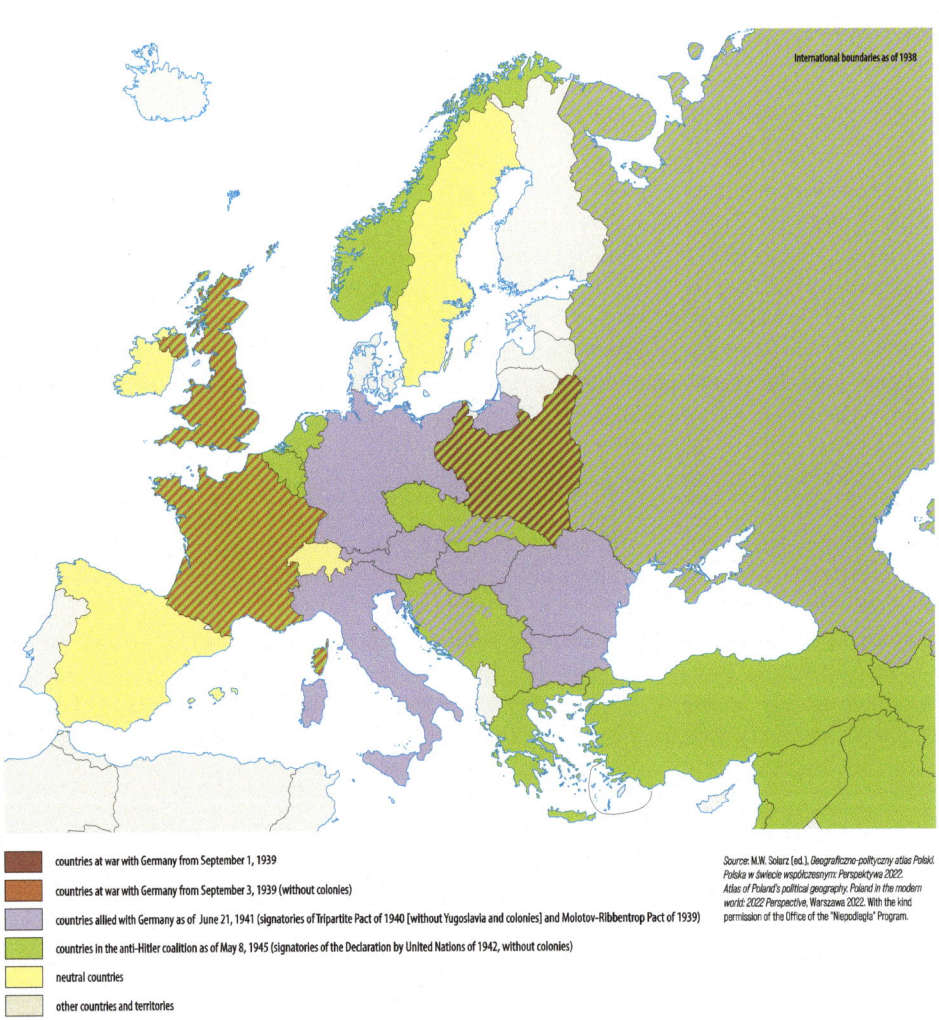

International boundaries as of 1938

countries at war with Germany from September 1, 1939

countries at war with Germany from September 3, 1939 (without colonies)

countries allied with Germany as of  June 21, 1941 (signatories of Tripartite Pact of 1940 [without Yugoslavia and colonies] and Molotov–Ribbentrop Pact of 1939)

countries in the anti-Hitler coalition as of May 8, 1945 (signatories of the Declaration by United Nations of 1942, without colonies)

neutral countries

other countries and territories

*Source: M.W. Solarz (ed.), Geograficzno-polityczny atlas Polski. Polska w świecie współczesnym: Perspektywa 2022. Atlas of Poland's political geography. Poland in the modern world: 2022 Perspective, Warszawa 2022. With the kind permission of the Office of the "Niepodległa" Program.*

*Figure 10.1* Europe in the face of the Third Reich

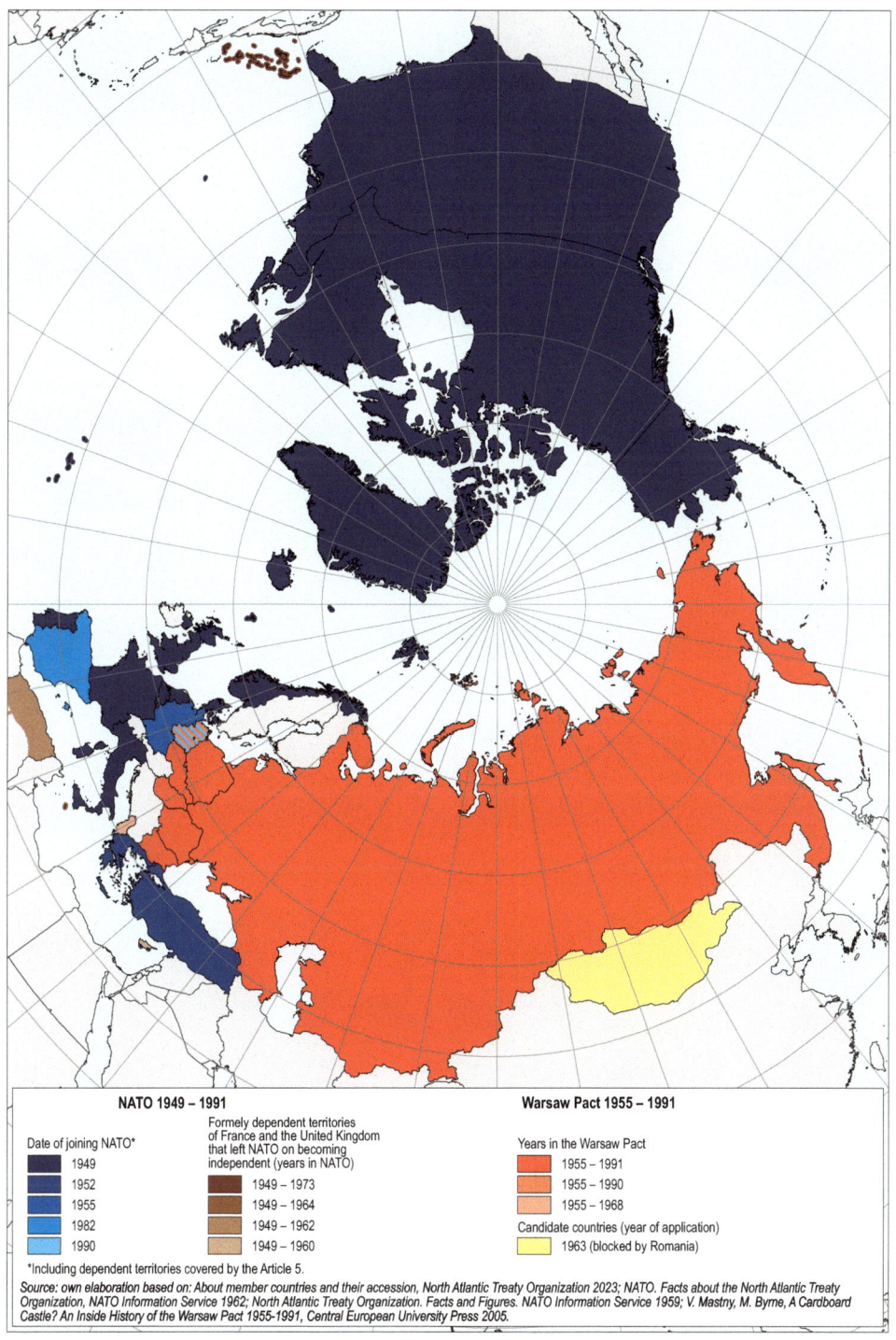

| NATO 1949 – 1991 | Formely dependent territories of France and the United Kingdom that left NATO on becoming independent (years in NATO) | Warsaw Pact 1955 – 1991 |
|---|---|---|

Date of joining NATO*

- 1949
- 1952
- 1955
- 1982
- 1990

- 1949 – 1973
- 1949 – 1964
- 1949 – 1962
- 1949 – 1960

Years in the Warsaw Pact

- 1955 – 1991
- 1955 – 1990
- 1955 – 1968

Candidate countries (year of application)

- 1963 (blocked by Romania)

*Including dependent territories covered by the Article 5.

*Source: own elaboration based on: About member countries and their accession, North Atlantic Treaty Organization 2023; NATO. Facts about the North Atlantic Treaty Organization, NATO Information Service 1962; North Atlantic Treaty Organization. Facts and Figures. NATO Information Service 1959; V. Mastny, M. Byrne, A Cardboard Castle? An Inside History of the Warsaw Pact 1955-1991, Central European University Press 2005.*

*Figure 10.2* Western and Eastern blocs until the end of the Cold War: NATO, 1949–1991, and the Warsaw Pact, 1955–1991

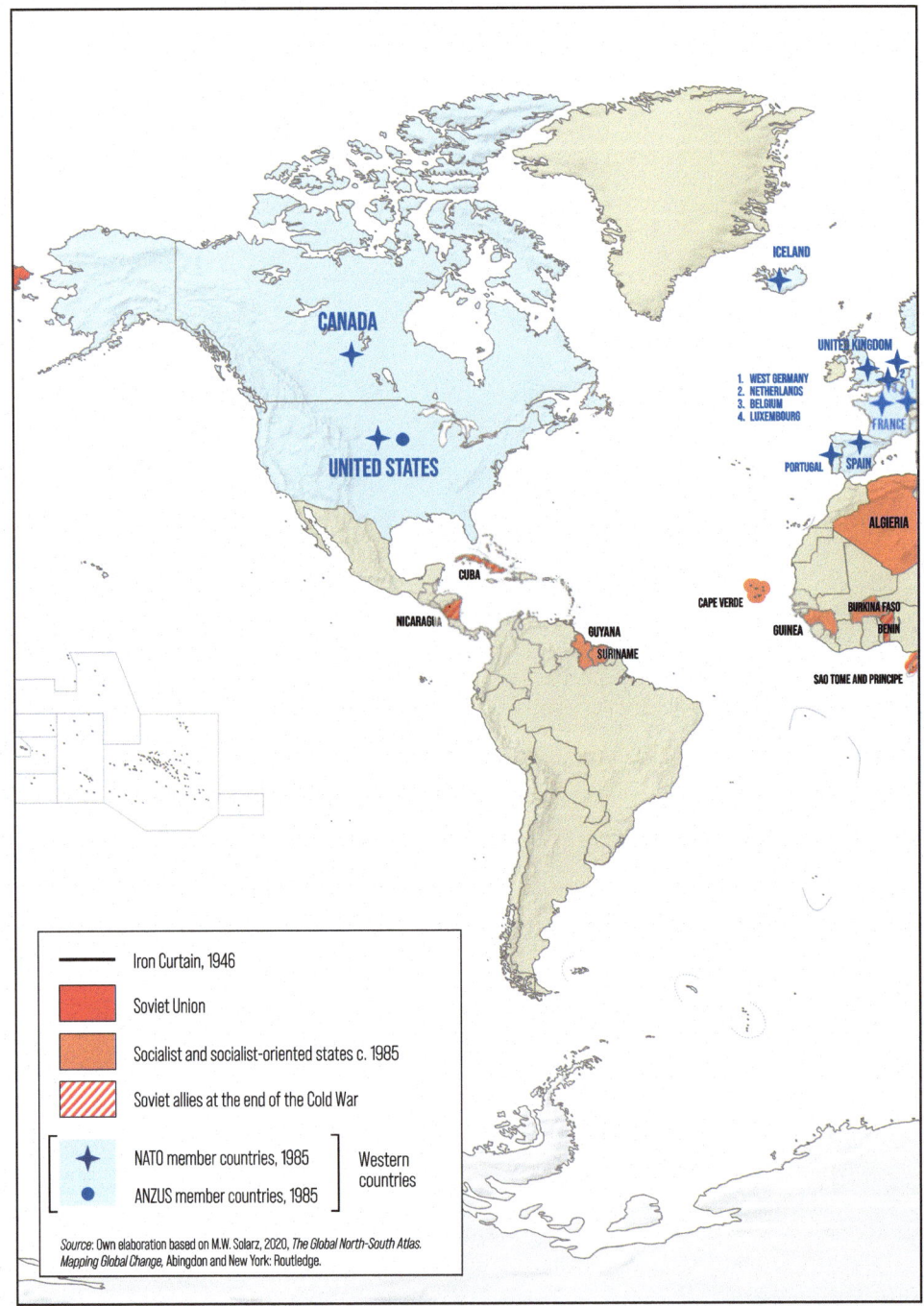

*Figure 10.3* The West vs the East during the Cold War

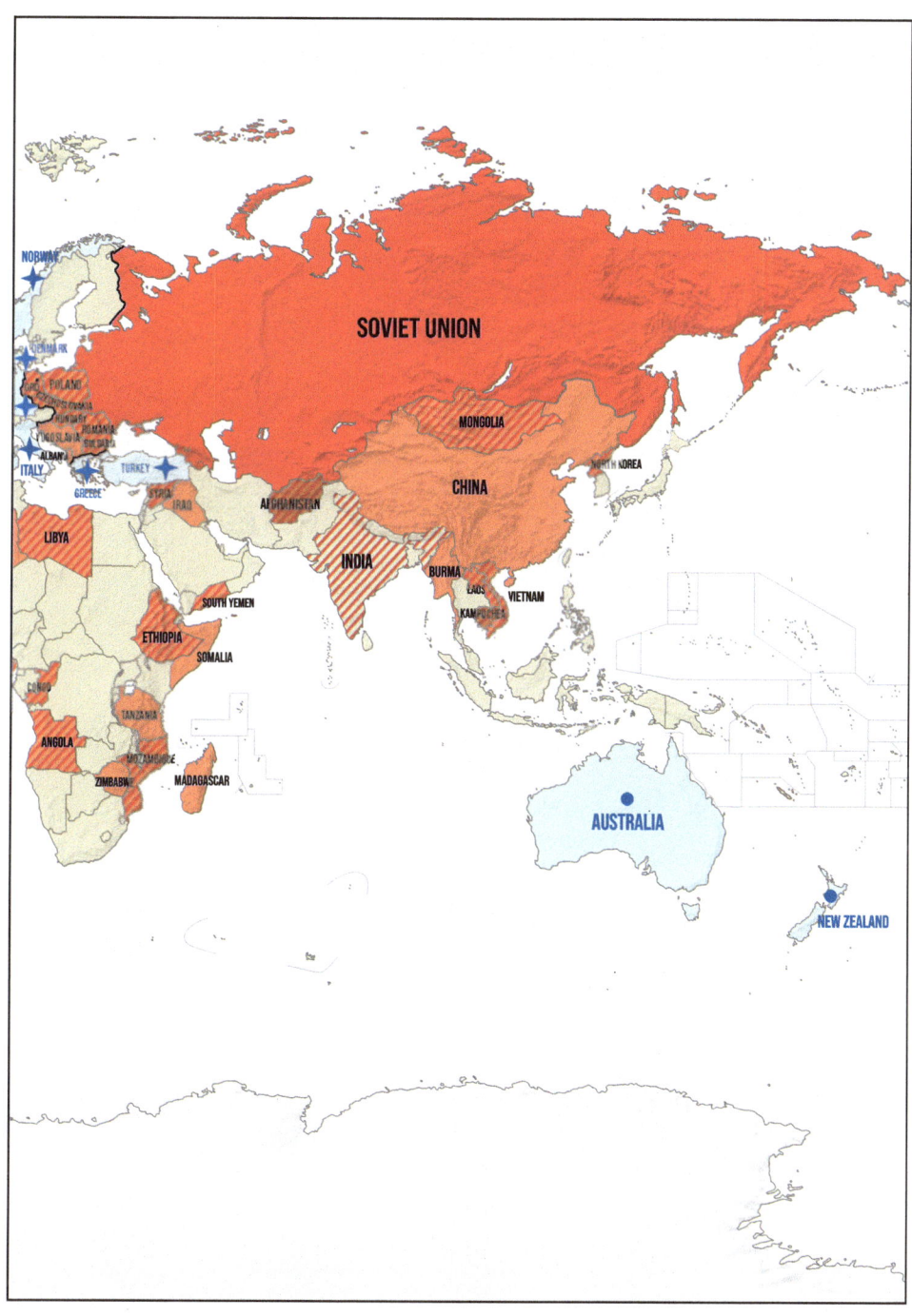

*Figure 10.3* (Continued)

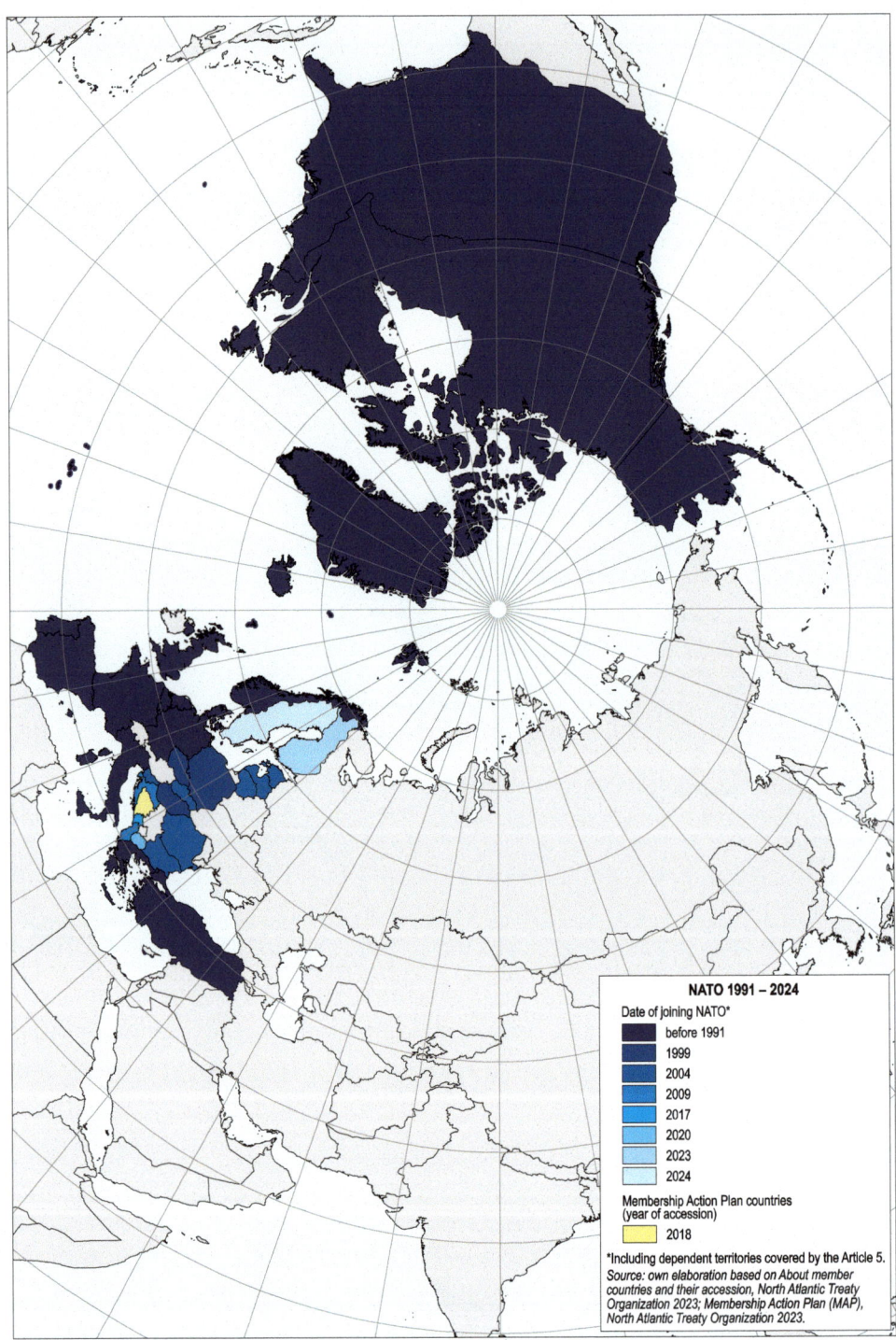

*Figure 10.4* NATO after the Cold War: 1991–2024

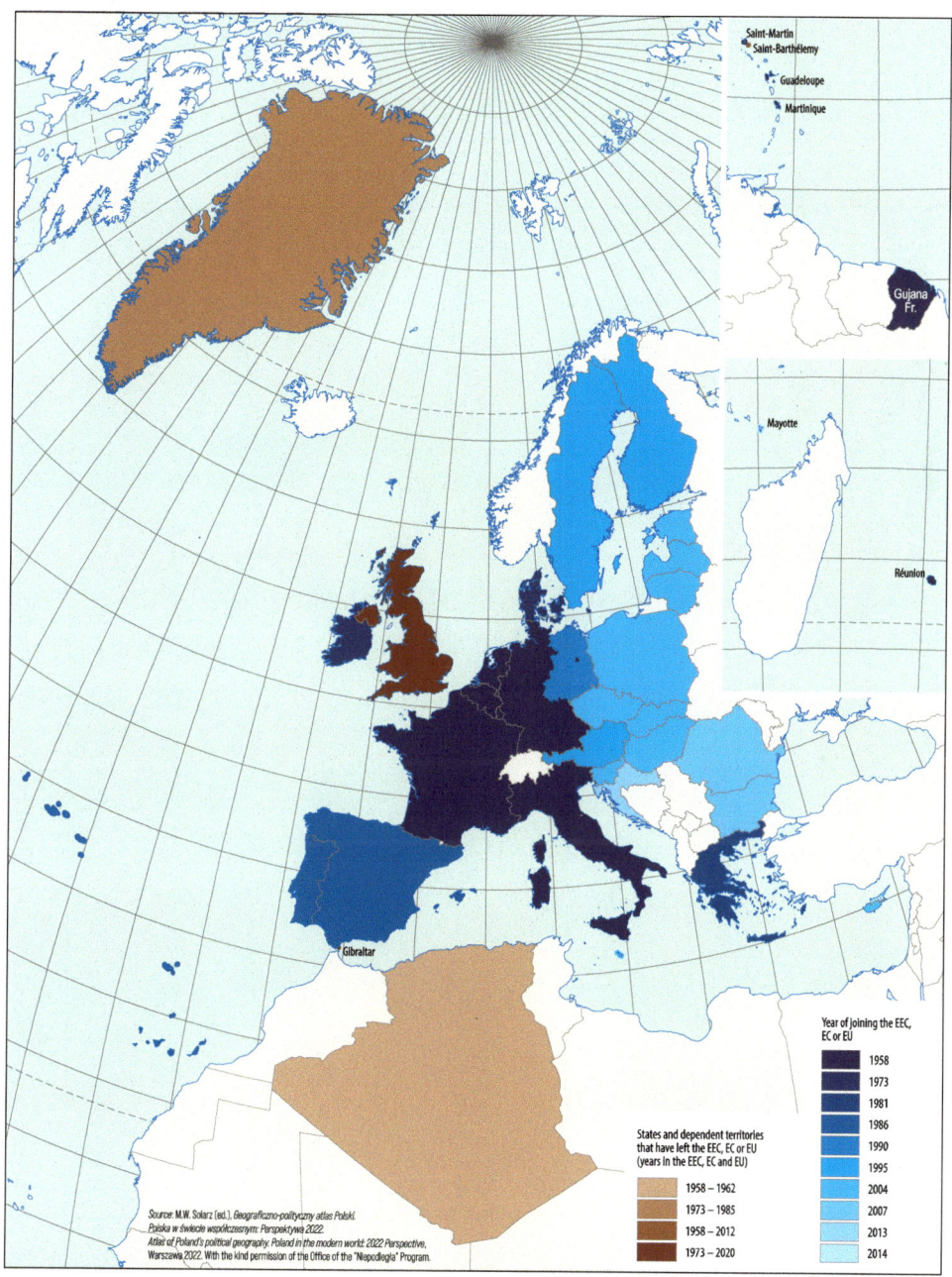

*Figure 10.5* The European Union – past and present

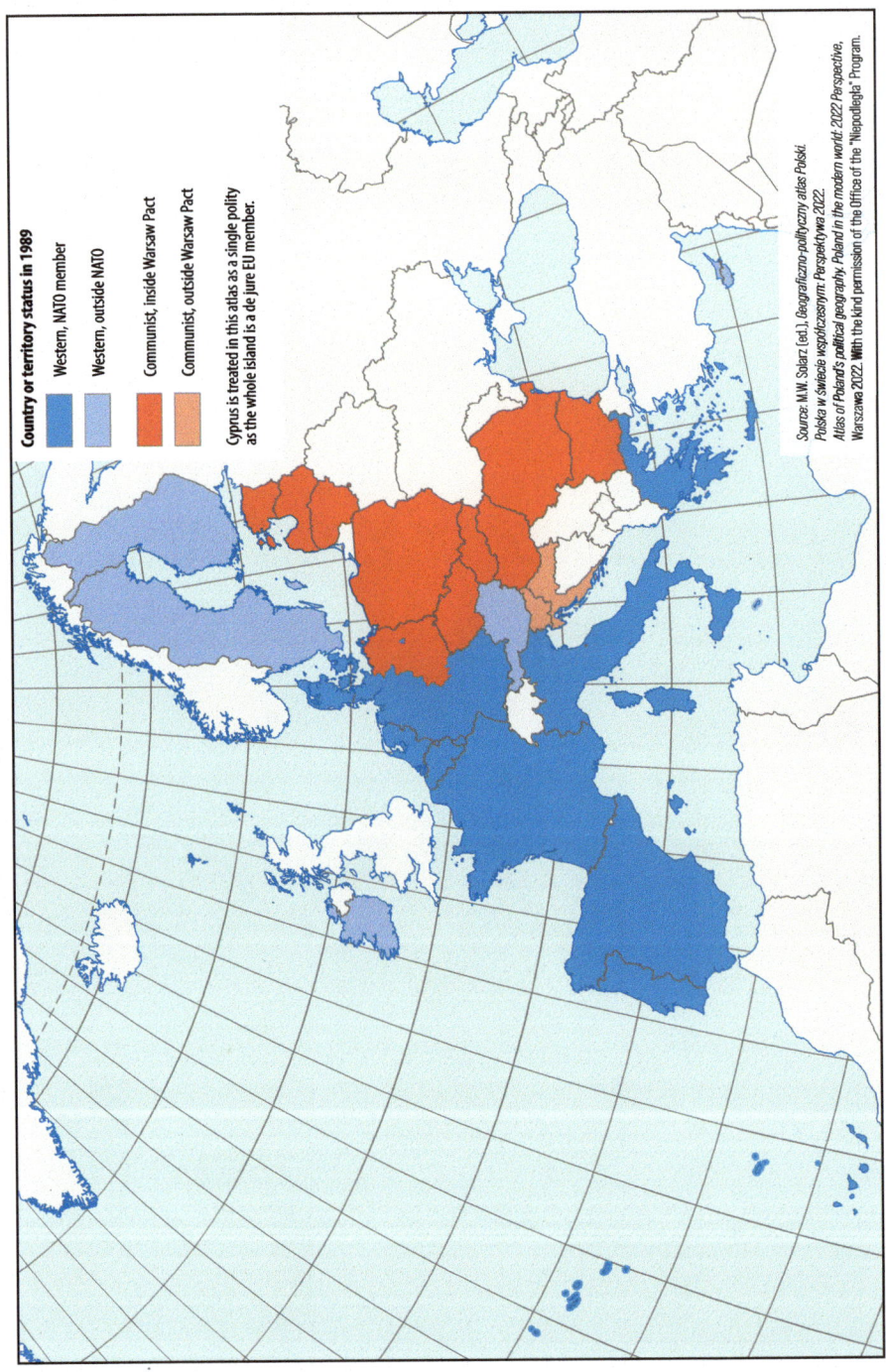

*Figure 10.6* The European Union – the geopolitical position of member states prior to 1989

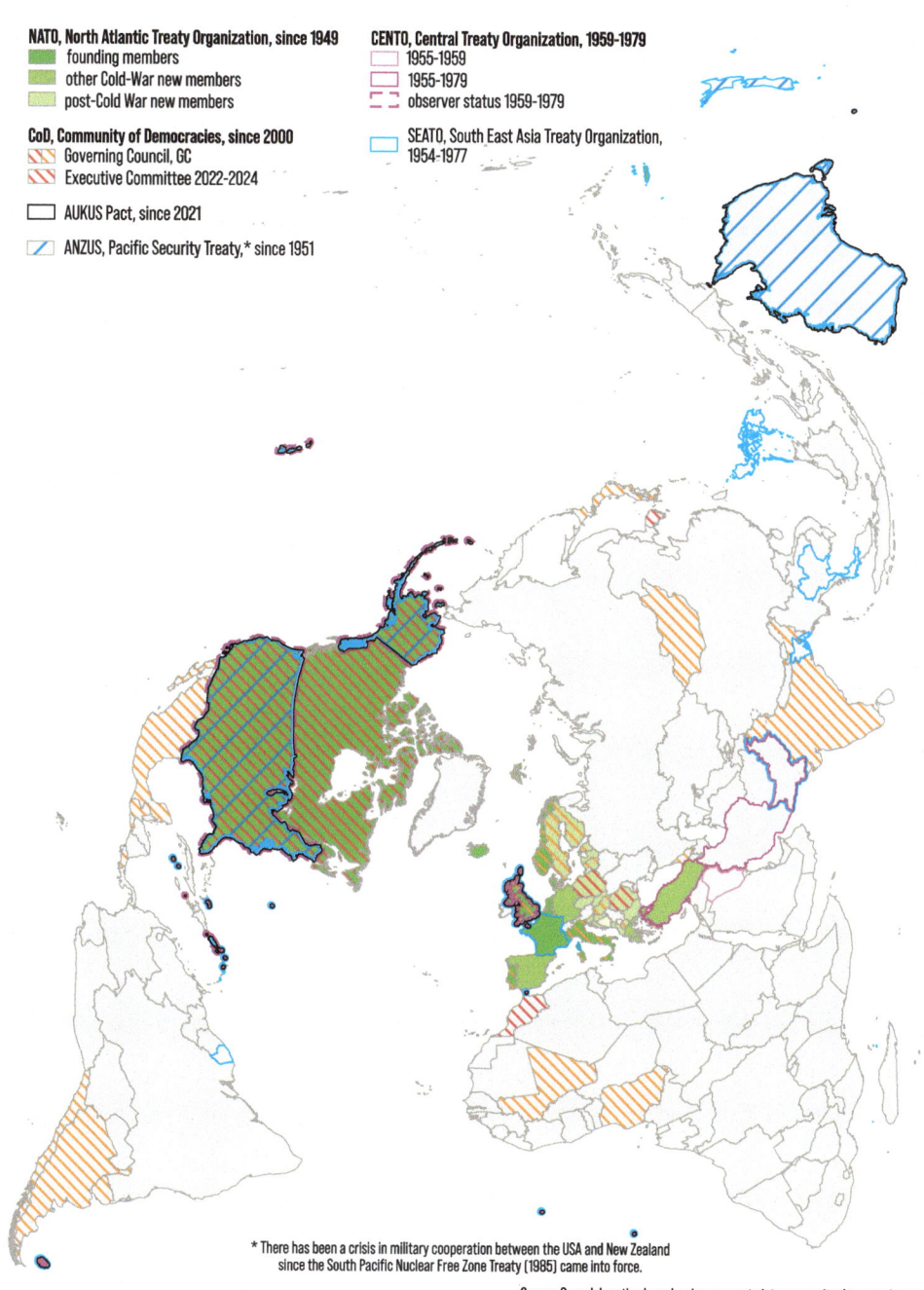

NATO, North Atlantic Treaty Organization, since 1949
founding members
other Cold-War new members
post-Cold War new members

CoD, Community of Democracies, since 2000
Governing Council, GC
Executive Committee 2022-2024

AUKUS Pact, since 2021

ANZUS, Pacific Security Treaty,* since 1951

CENTO, Central Treaty Organization, 1959-1979
1955-1959
1955-1979
observer status 1959-1979

SEATO, South East Asia Treaty Organization, 1954-1977

* There has been a crisis in military cooperation between the USA and New Zealand since the South Pacific Nuclear Free Zone Treaty (1985) came into force.

*Source:* Own elaboration based on i.a. www.nato.int; community-democracies.org

*Figure 10.7* The Global System of Alliances I: 1945–2024, Western system

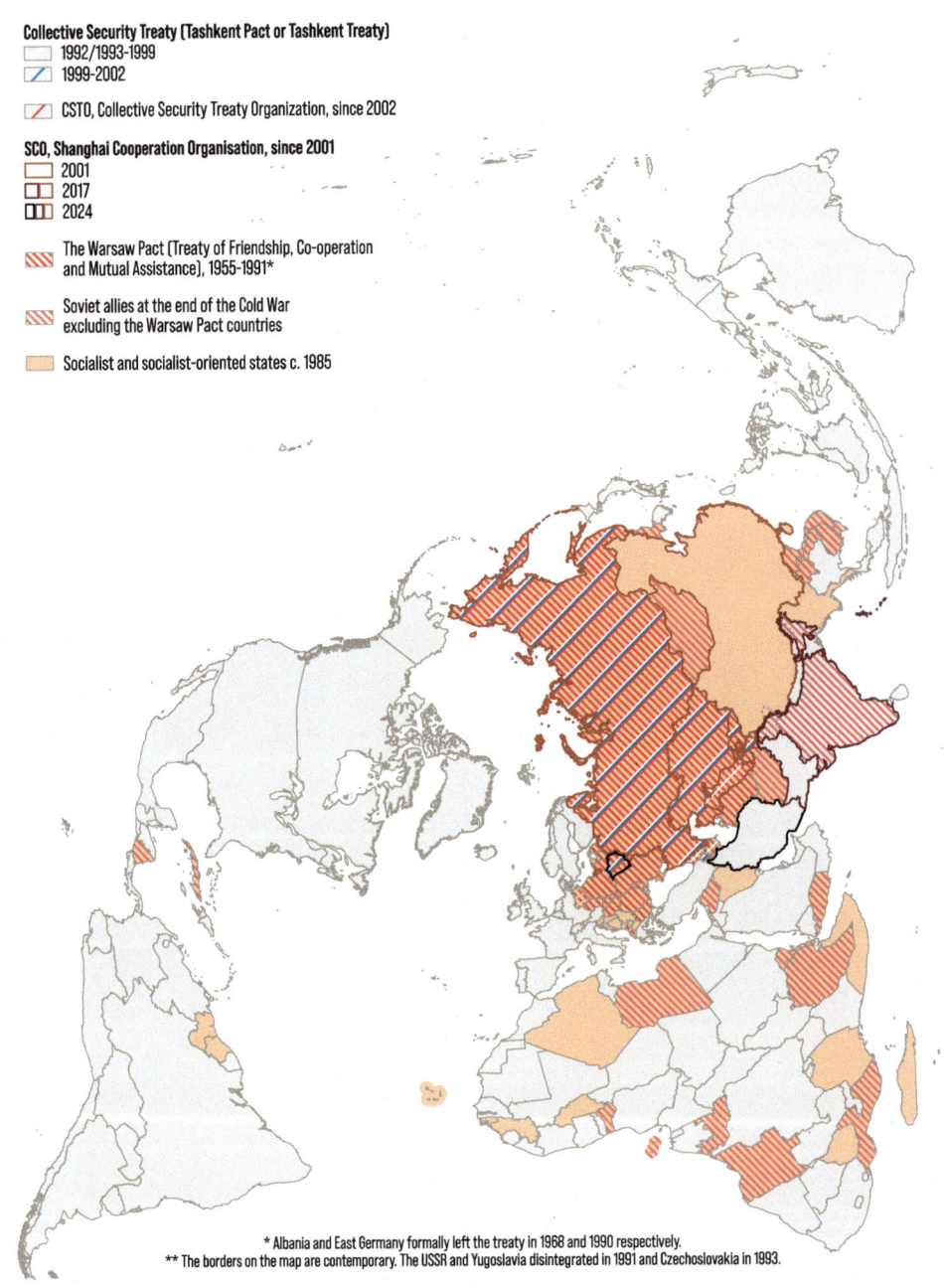

**Collective Security Treaty (Tashkent Pact or Tashkent Treaty)**
- 1992/1993-1999
- 1999-2002

CSTO, Collective Security Treaty Organization, since 2002

**SCO, Shanghai Cooperation Organisation, since 2001**
- 2001
- 2017
- 2024

The Warsaw Pact (Treaty of Friendship, Co-operation and Mutual Assistance), 1955-1991*

Soviet allies at the end of the Cold War excluding the Warsaw Pact countries

Socialist and socialist-oriented states c. 1985

* Albania and East Germany formally left the treaty in 1968 and 1990 respectively.
** The borders on the map are contemporary. The USSR and Yugoslavia disintegrated in 1991 and Czechoslovakia in 1993.

Source: Own elaboration based on i.a. M.W. Solarz, 2020, The Global North-South Atlas. Mapping Global Change, Abingdon and New York: Routledge.

*Figure 10.8* The Global System of Alliances II: 1945–2024, Eastern system

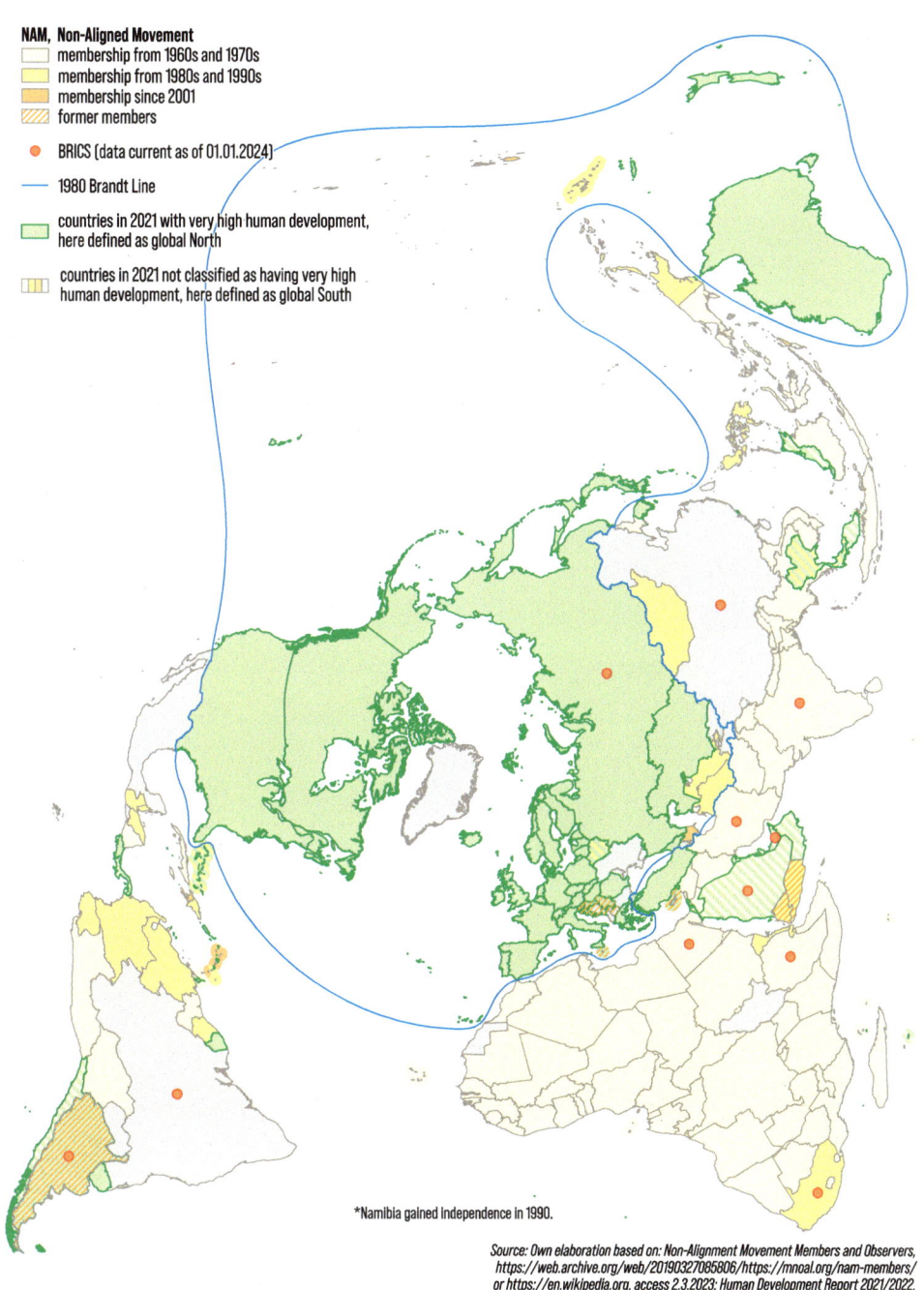

**NAM, Non-Aligned Movement**
- membership from 1960s and 1970s
- membership from 1980s and 1990s
- membership since 2001
- former members

- BRICS (data current as of 01.01.2024)

— 1980 Brandt Line

countries in 2021 with very high human development, here defined as global North

countries in 2021 not classified as having very high human development, here defined as global South

*Namibia gained independence in 1990.

*Source: Own elaboration based on: Non-Alignment Movement Members and Observers, https://web.archive.org/web/20190327085806/https://mnoal.org/nam-members/ or https://en.wikipedia.org, access 2.3.2023; Human Development Report 2021/2022.*

*Figure 10.9* The Global System of Alliances III: 1945–2024, Southern system

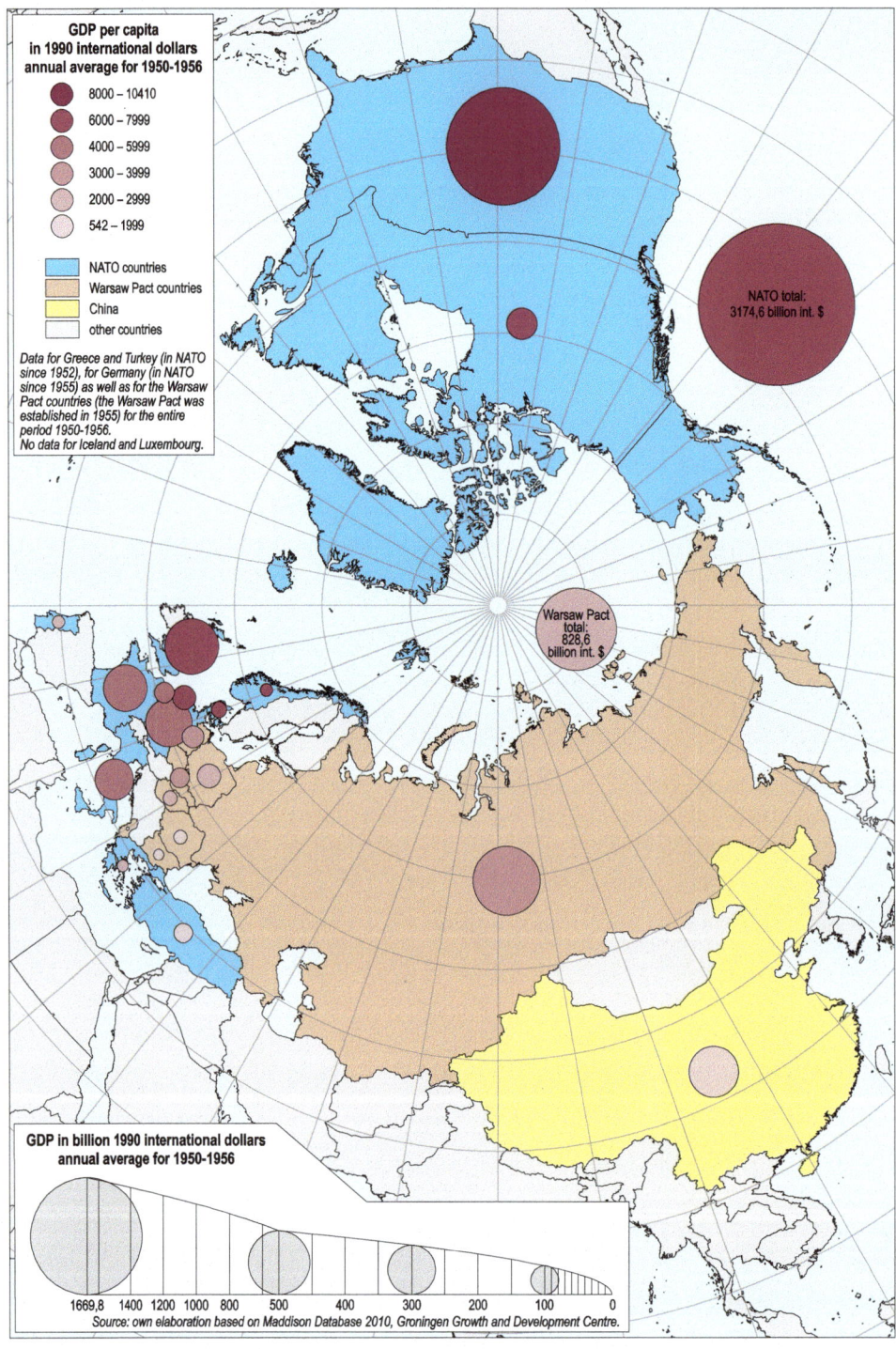

*Figure 10.10* GDP and per capita GDP: NATO vs the Warsaw Pact and the People's Republic of
China: 1950–1956

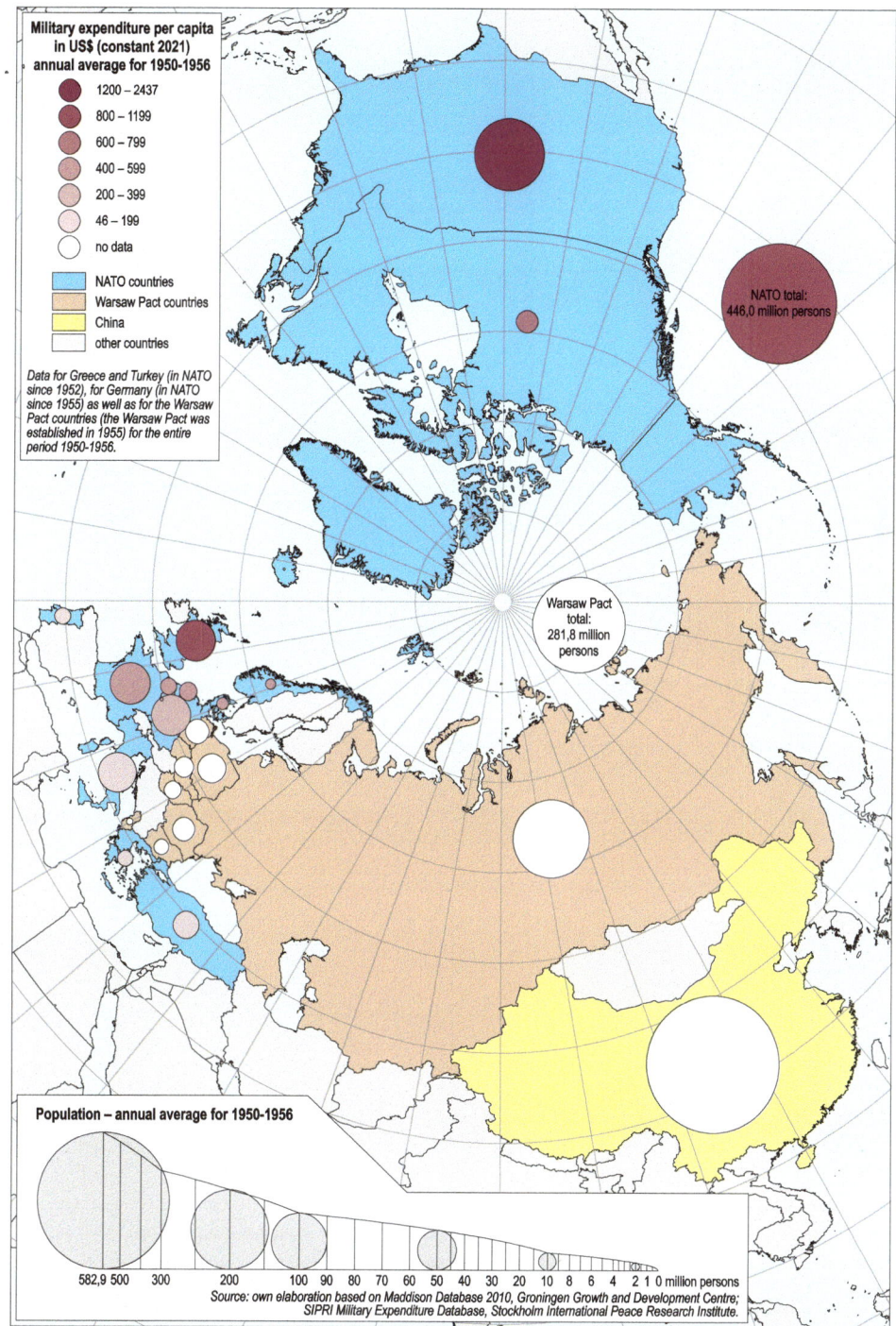

*Figure 10.11* Military expenditure per capita and population: NATO vs the Warsaw Pact and the People's Republic of China: 1950–1956

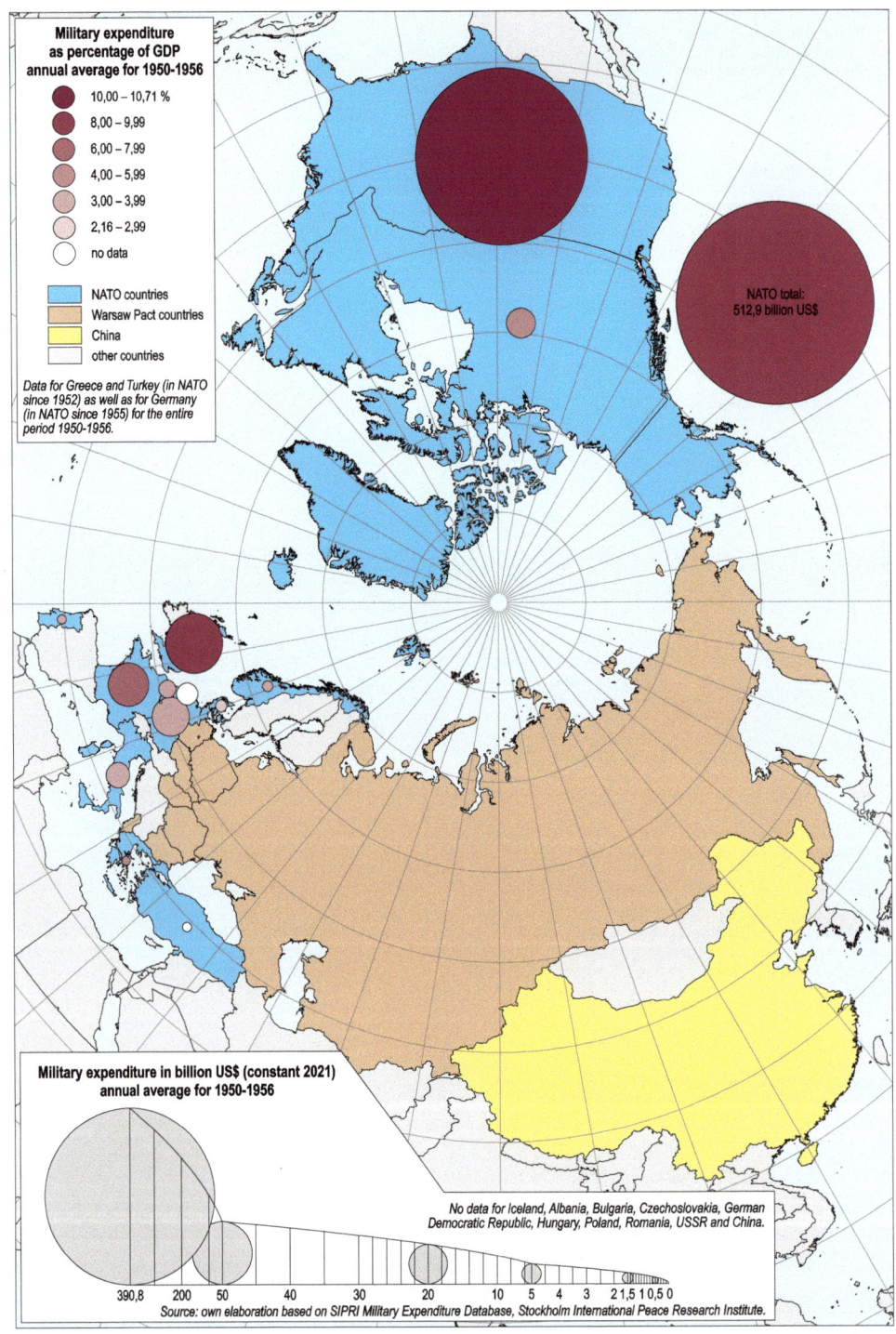

*Figure 10.12* Military expenditure as a percentage of GDP and in absolute terms: NATO vs the Warsaw Pact and the People's Republic of China: 1950–1956

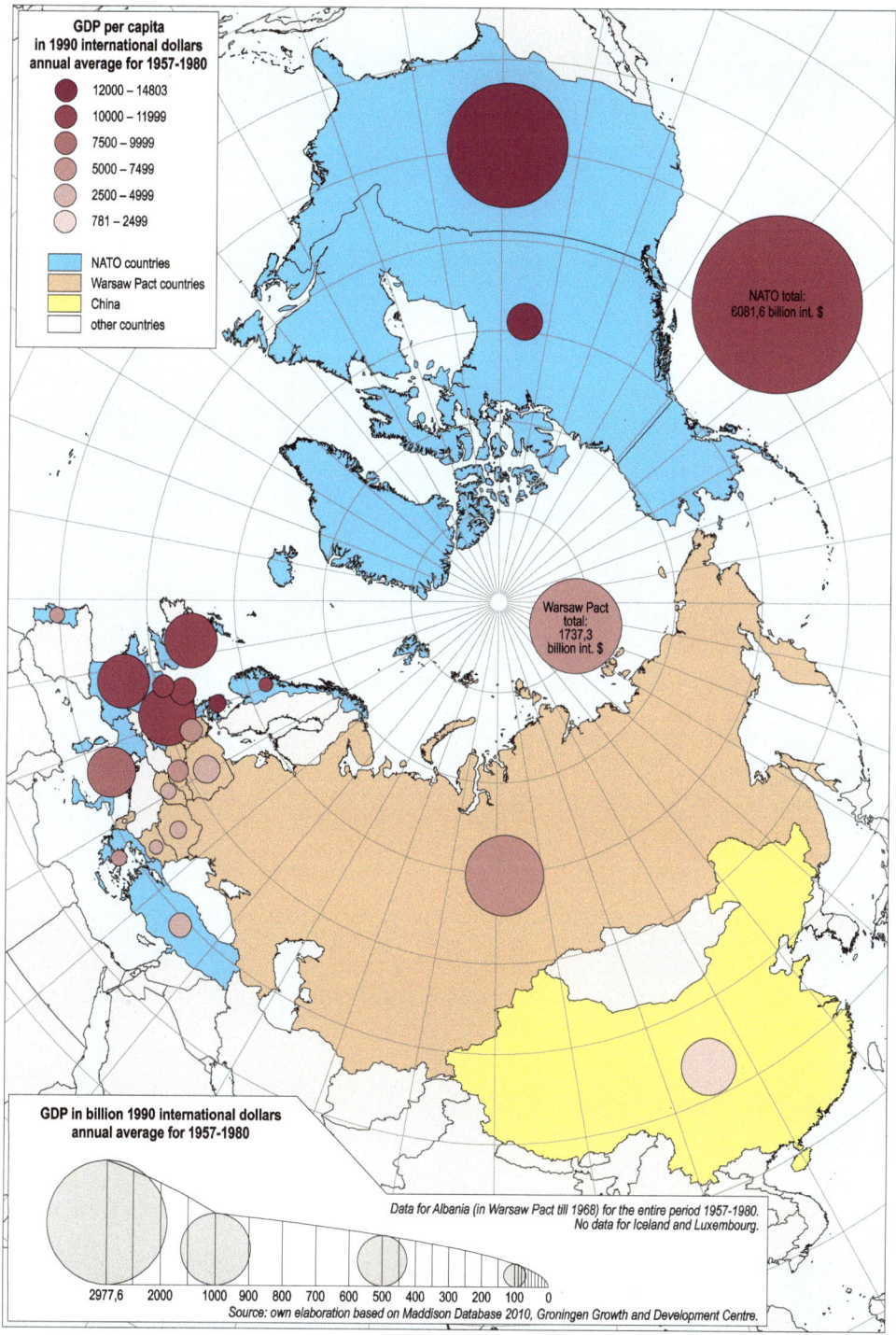

GDP per capita
in 1990 international dollars
annual average for 1957-1980

- 12000 – 14803
- 10000 – 11999
- 7500 – 9999
- 5000 – 7499
- 2500 – 4999
- 781 – 2499

NATO countries
Warsaw Pact countries
China
other countries

NATO total:
6081,6 billion int. $

Warsaw Pact
total:
1737,3
billion int. $

GDP in billion 1990 international dollars
annual average for 1957-1980

Data for Albania (in Warsaw Pact till 1968) for the entire period 1957-1980.
No data for Iceland and Luxembourg.

2977,6  2000  1000 900 800 700 600 500 400 300 200 100  0

Source: own elaboration based on Maddison Database 2010, Groningen Growth and Development Centre.

*Figure 10.13* GDP and per capita GDP: NATO vs the Warsaw Pact and the People's Republic of
China: 1957–1980

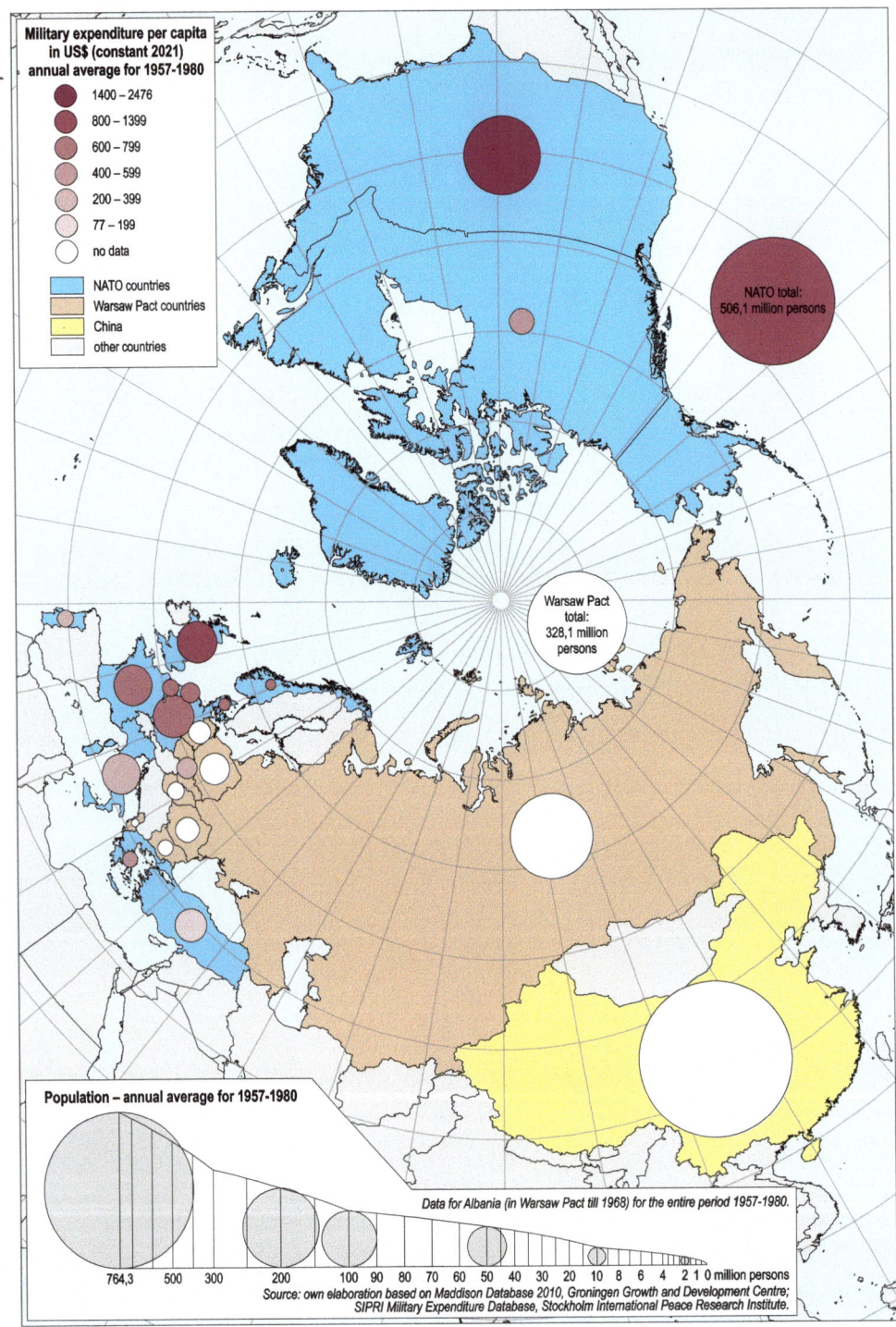

*Figure 10.14* Military expenditure per capita and population: NATO vs the Warsaw Pact and the People's Republic of China: 1957–1980

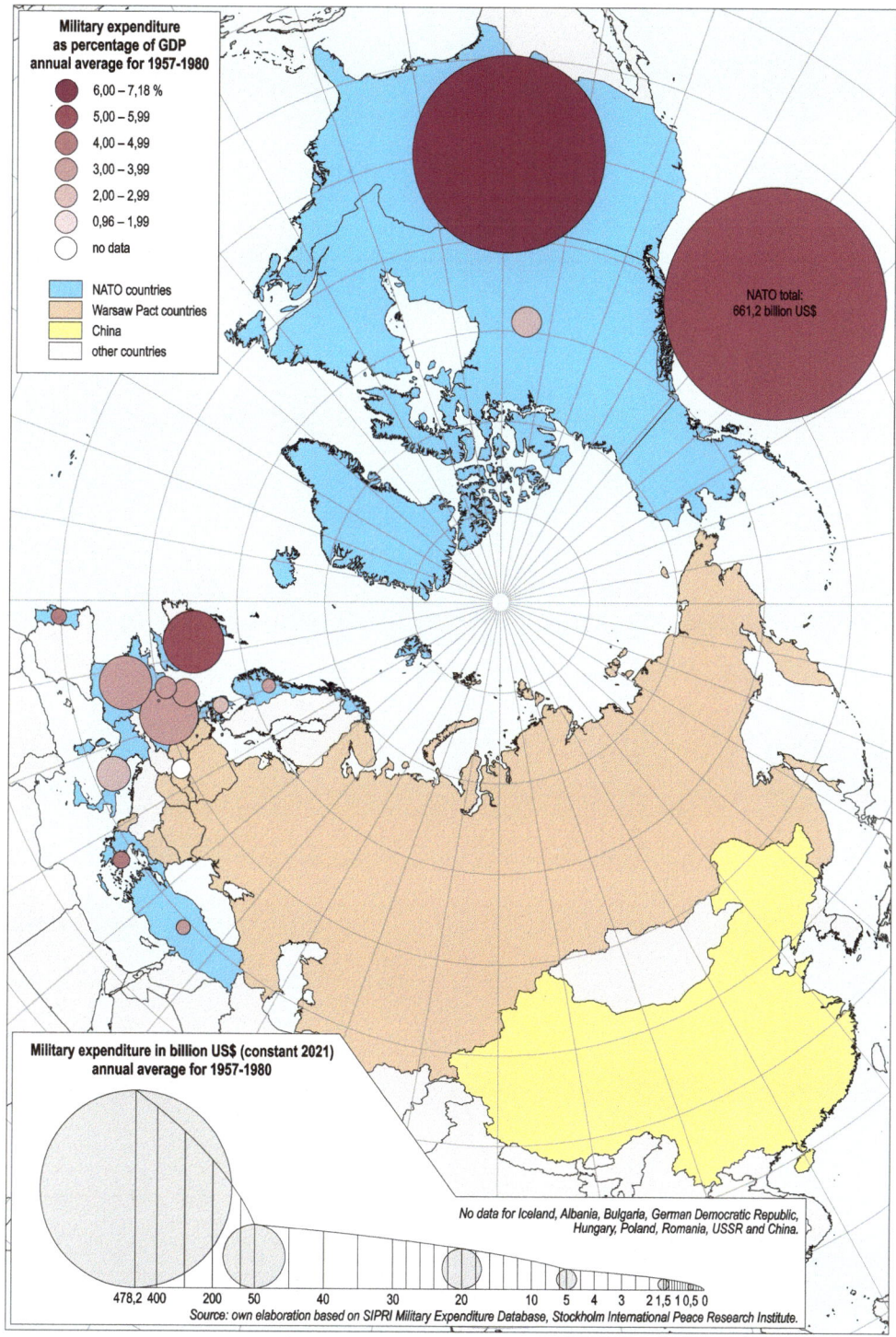

*Figure 10.15* Military expenditure as a percentage of GDP and in absolute terms: NATO vs the Warsaw Pact and the People's Republic of China: 1957–1980

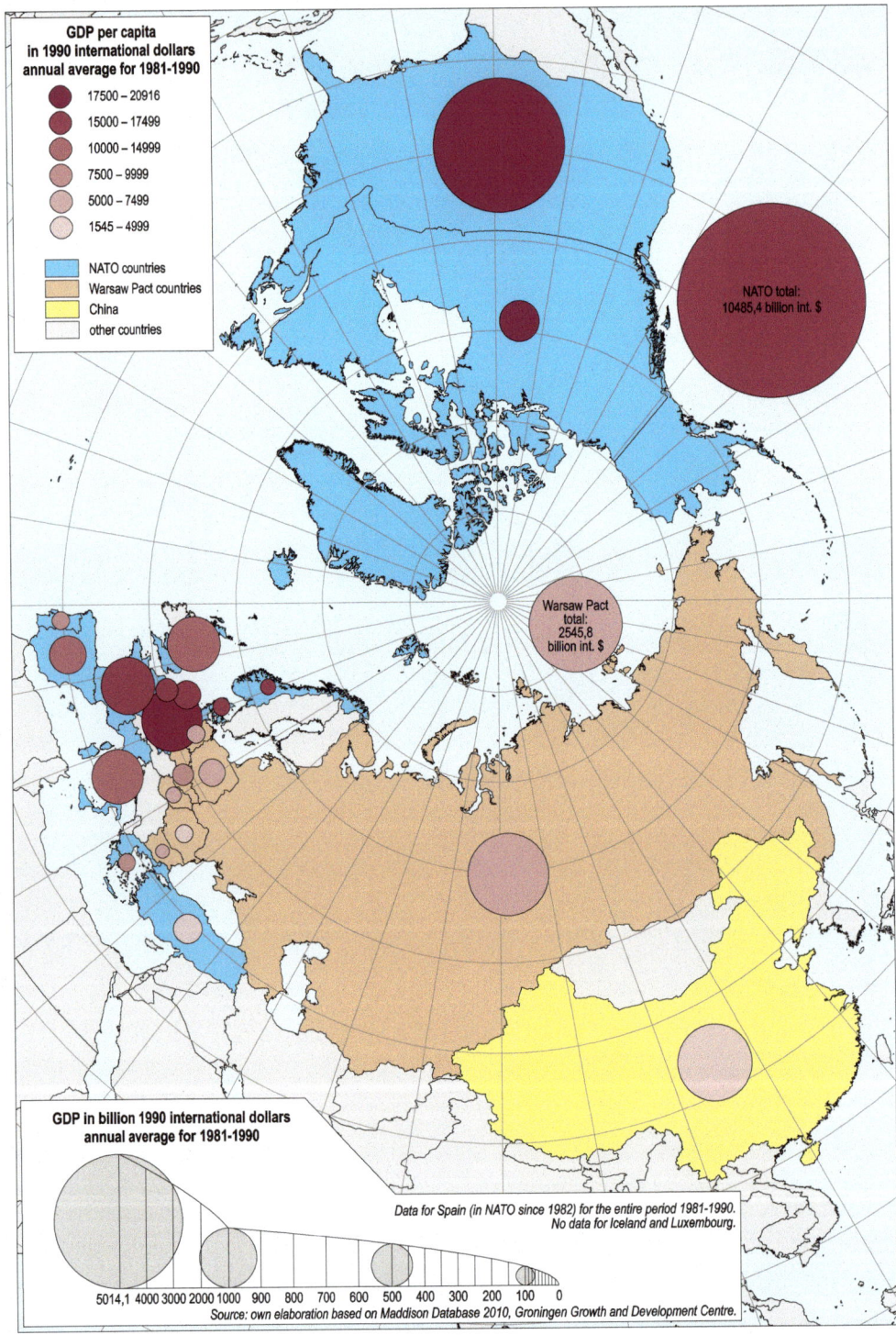

*Figure 10.16* GDP and per capita GDP: NATO vs the Warsaw Pact and the People's Republic of China: 1981–1990

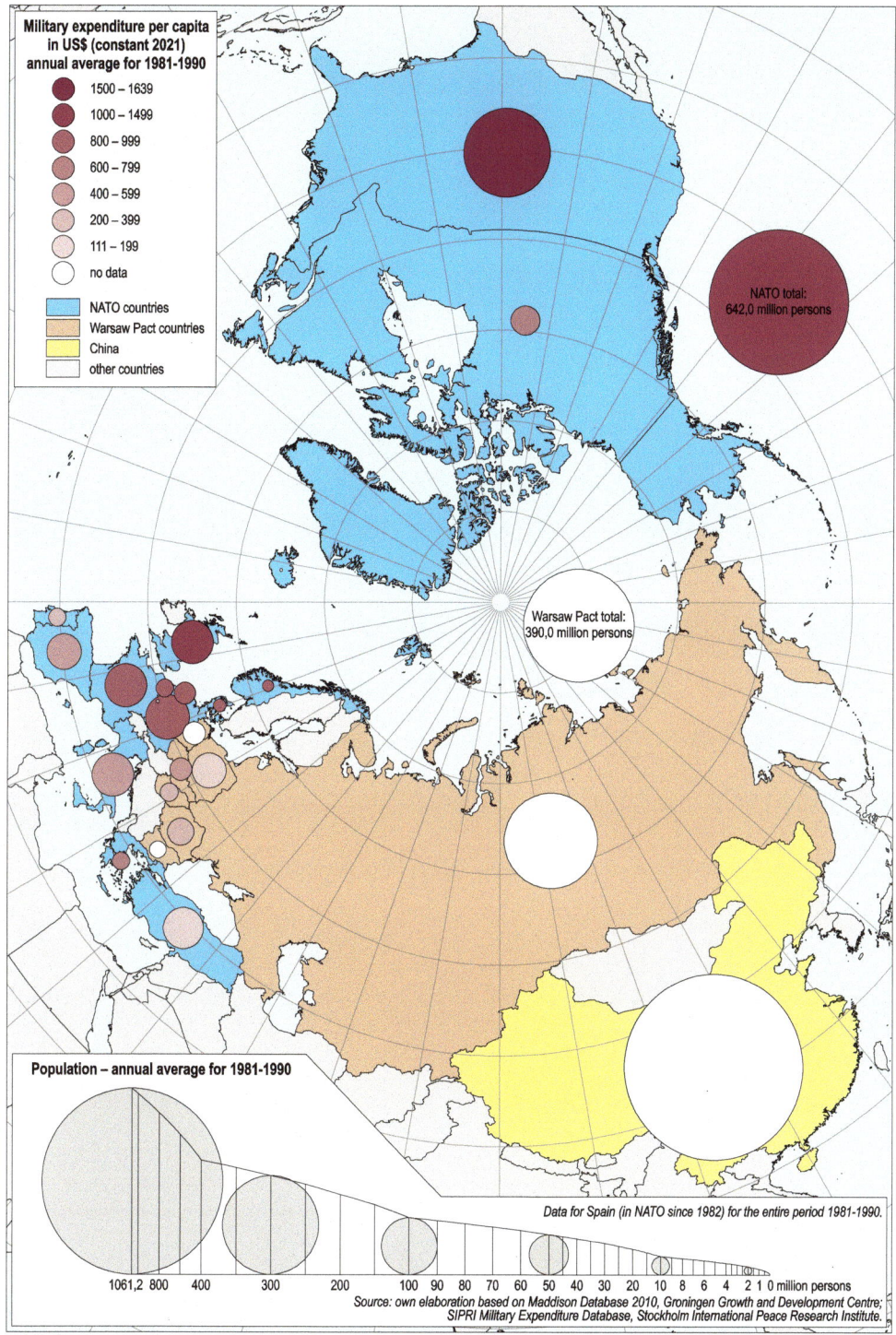

*Figure 10.17* Military expenditure per capita and population: NATO vs the Warsaw Pact and the People's Republic of China: 1981–1990

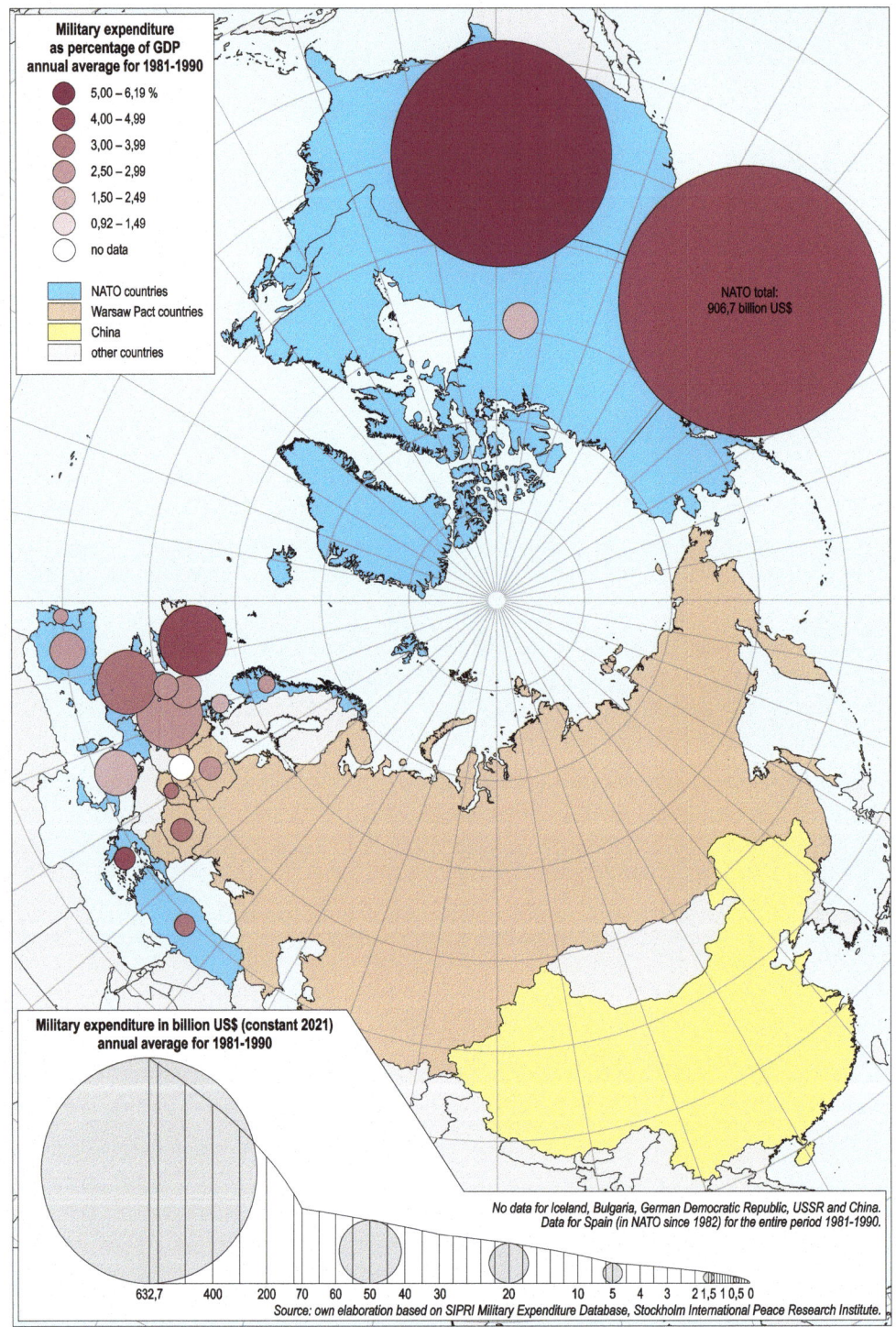

*Figure 10.18* Military expenditure as a percentage of GDP and in absolute terms: NATO vs the Warsaw Pact and the People's Republic of China: 1981–1990

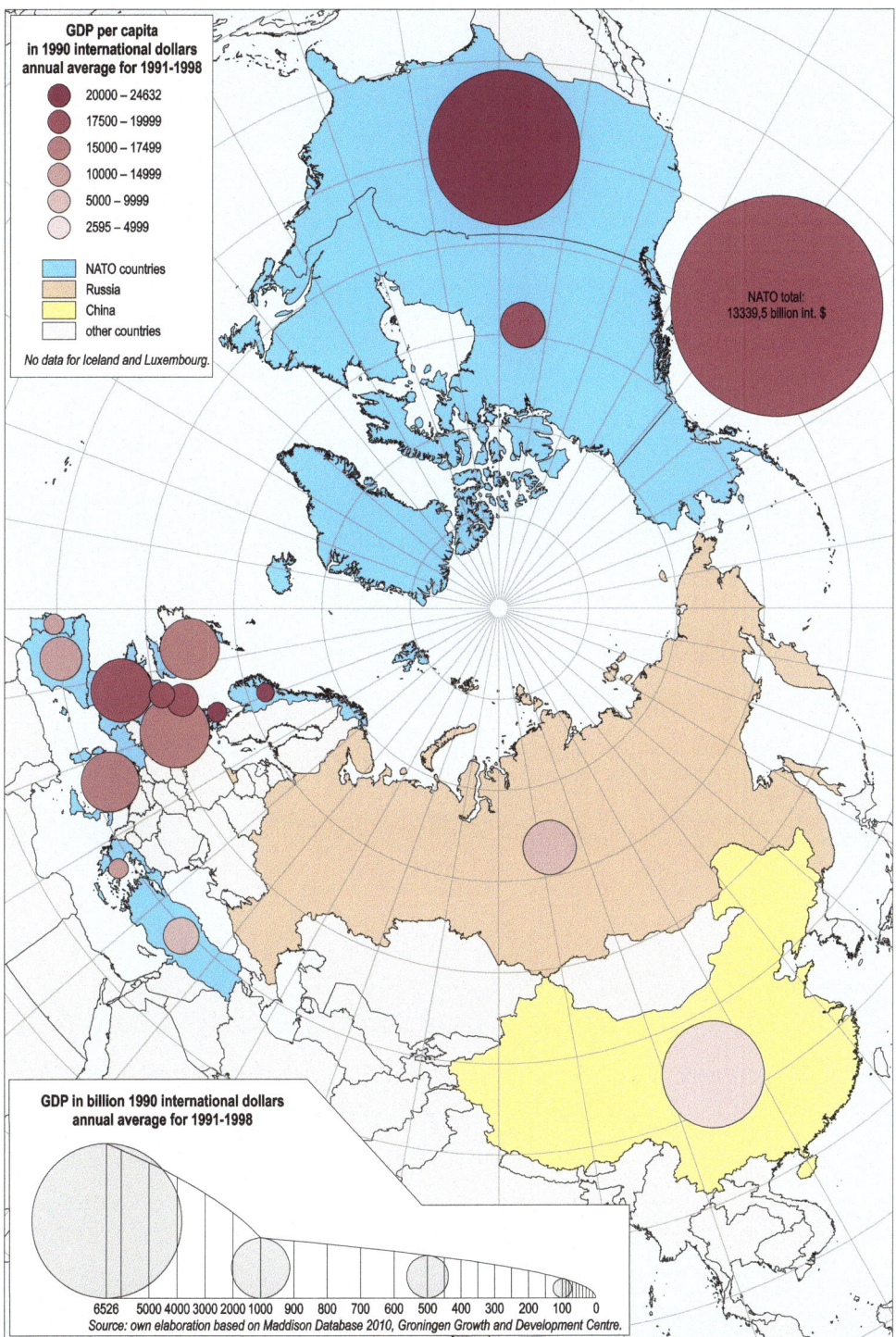

*Figure 10.19* GDP and per capita GDP: NATO vs Russia and the People's Republic of China:
1991–1998

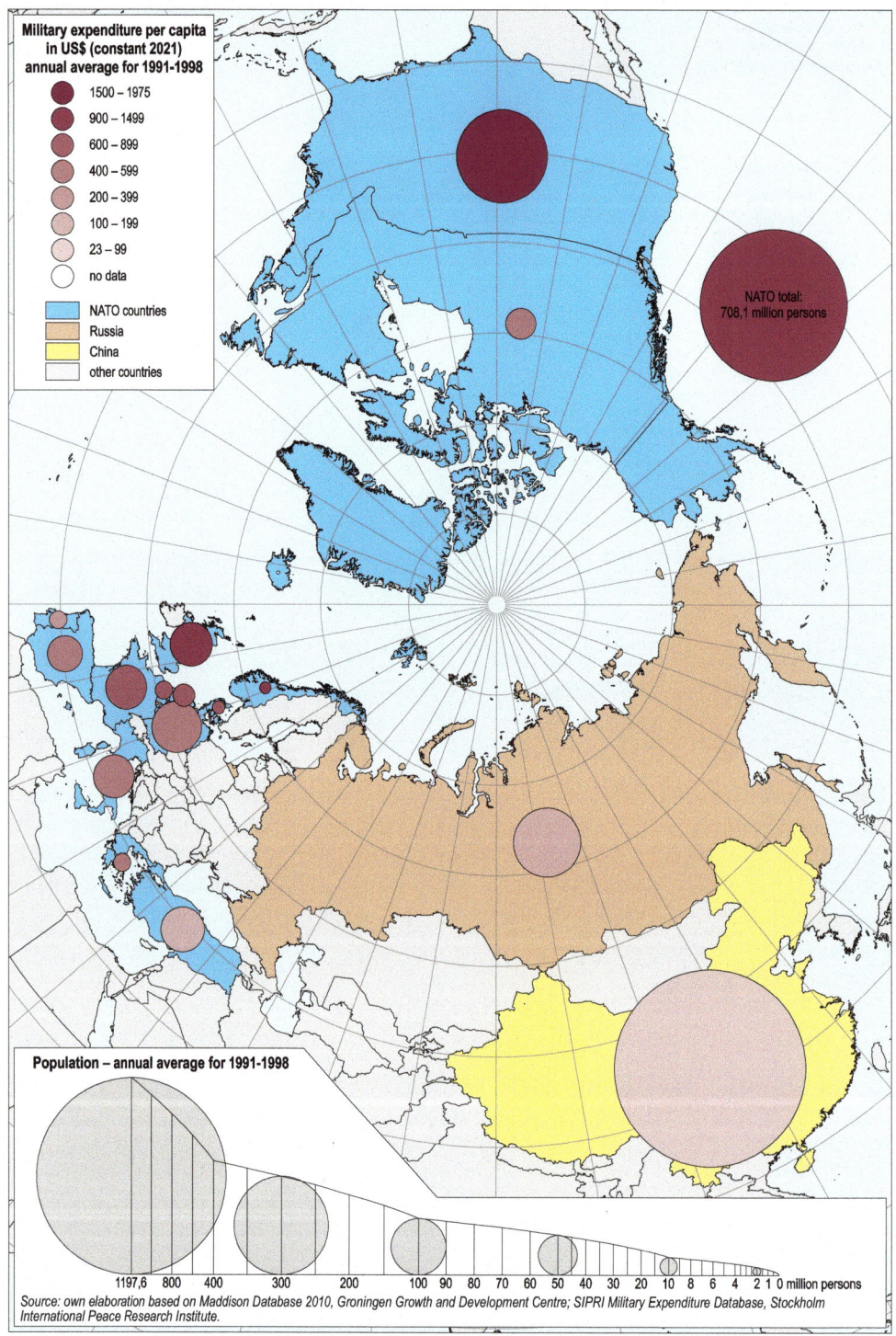

*Figure 10.20* Military expenditure per capita and population: NATO vs Russia and the People's Republic of China: 1991–1998

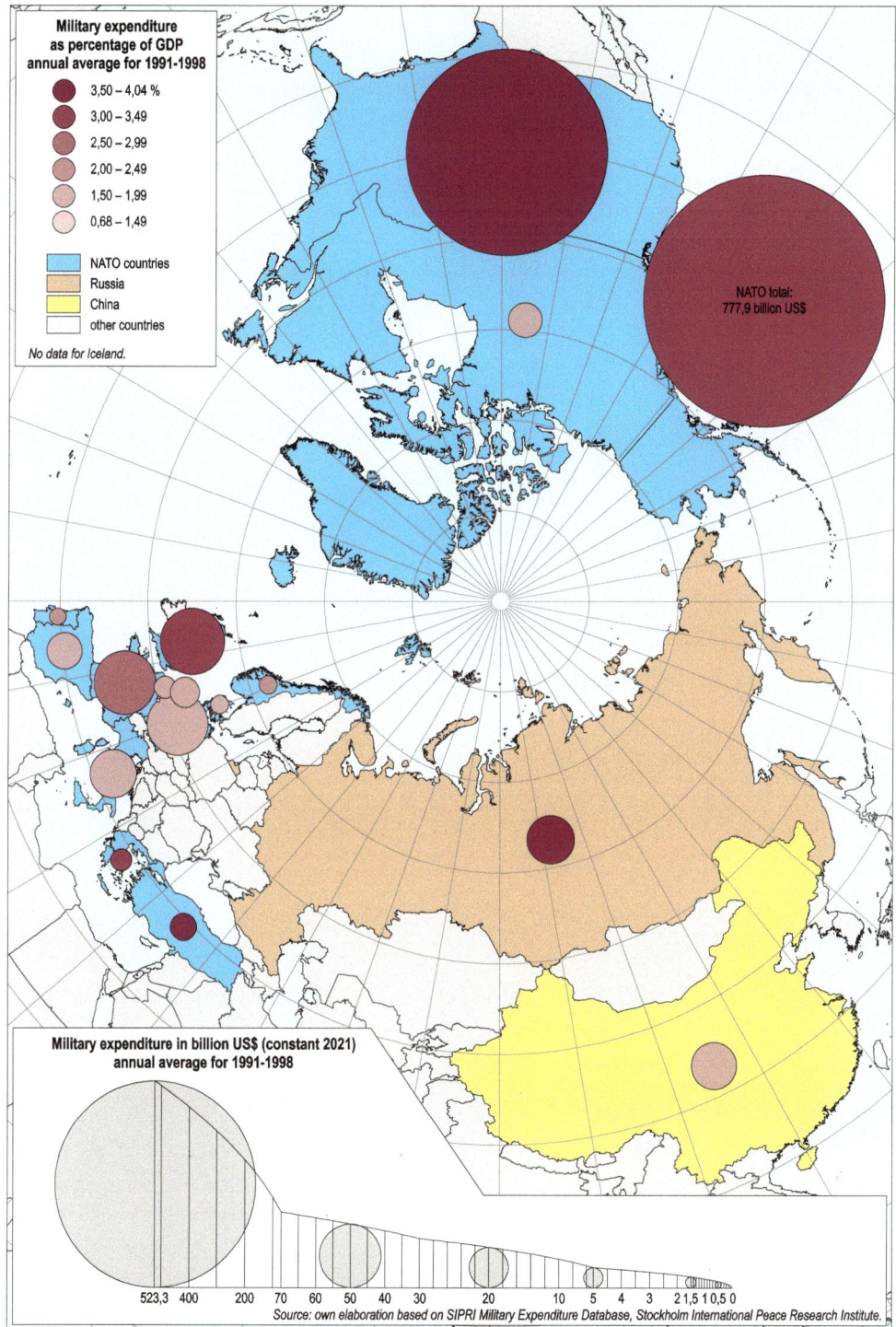

*Figure 10.21* Military expenditure as a percentage of GDP and in absolute terms: NATO vs Russia and the People's Republic of China: 1991–1998

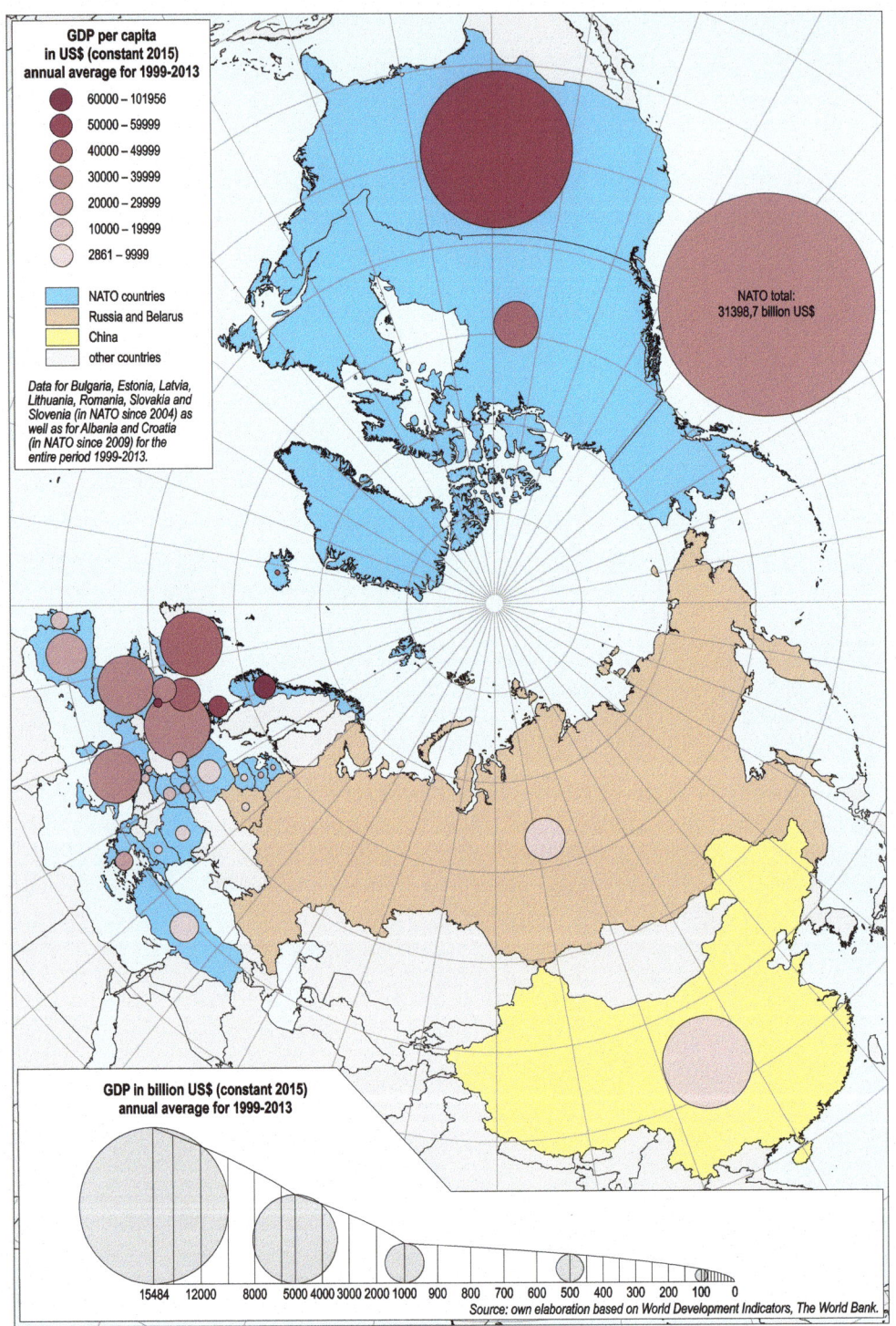

*Figure 10.22* GDP and per capita GDP: NATO vs Russia, Belarus, and the People's Republic of China: 1999–2013

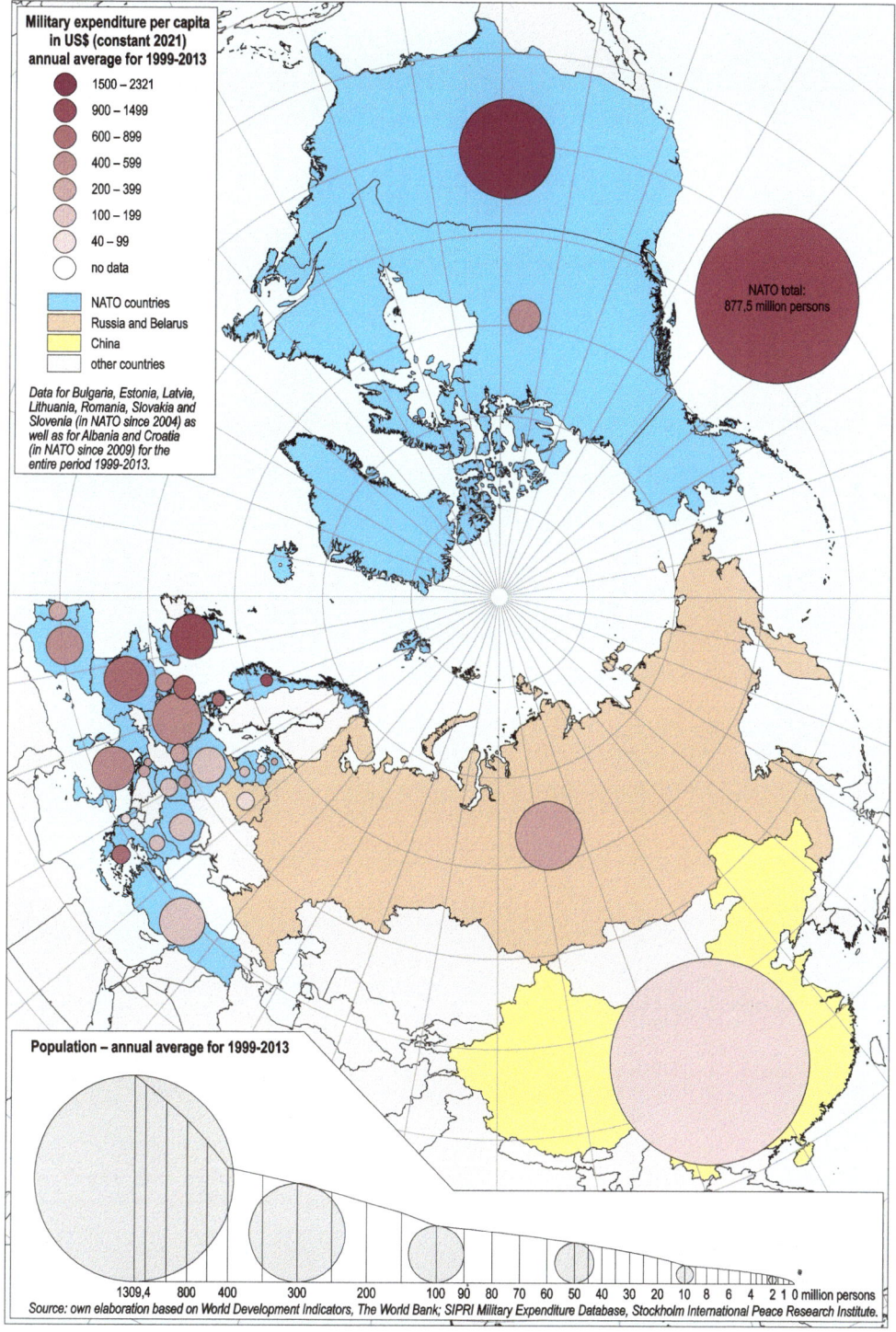

**Military expenditure per capita in US$ (constant 2021) annual average for 1999-2013**

1500 – 2321
900 – 1499
600 – 899
400 – 599
200 – 399
100 – 199
40 – 99
no data

NATO countries
Russia and Belarus
China
other countries

Data for Bulgaria, Estonia, Latvia, Lithuania, Romania, Slovakia and Slovenia (in NATO since 2004) as well as for Albania and Croatia (in NATO since 2009) for the entire period 1999-2013.

NATO total: 877,5 million persons

**Population – annual average for 1999-2013**

1309,4   800   400   300   200   100 90 80 70 60 50 40 30 20 10 8 6 4 2 1 0 million persons

Source: own elaboration based on World Development Indicators, The World Bank; SIPRI Military Expenditure Database, Stockholm International Peace Research Institute.

*Figure 10.23* Military expenditure per capita and population: NATO vs Russia, Belarus, and the People's Republic of China: 1999–2013

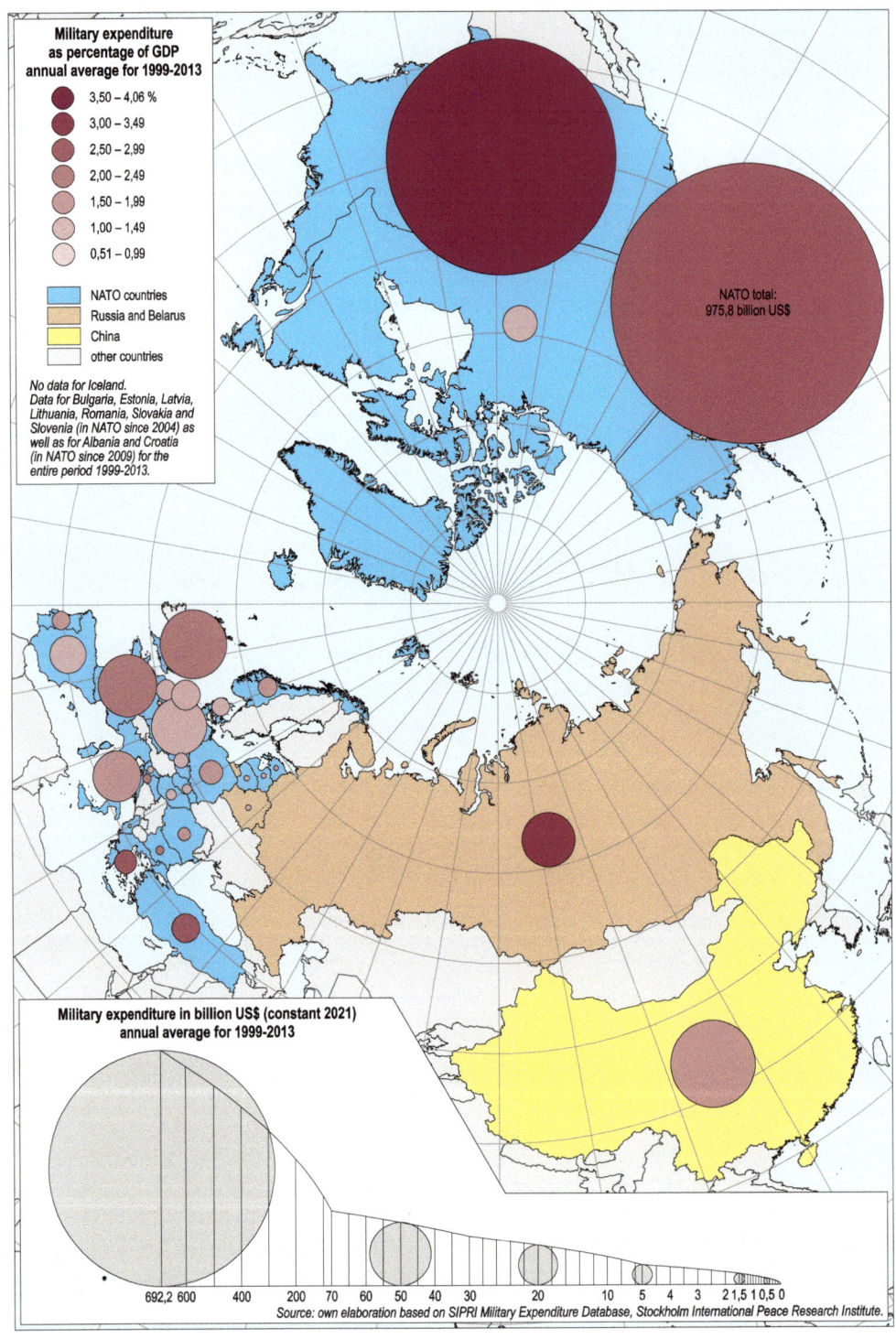

*Figure 10.24* Military expenditure as a percentage of GDP and in absolute terms: NATO vs Russia, Belarus, and the People's Republic of China: 1999–2013

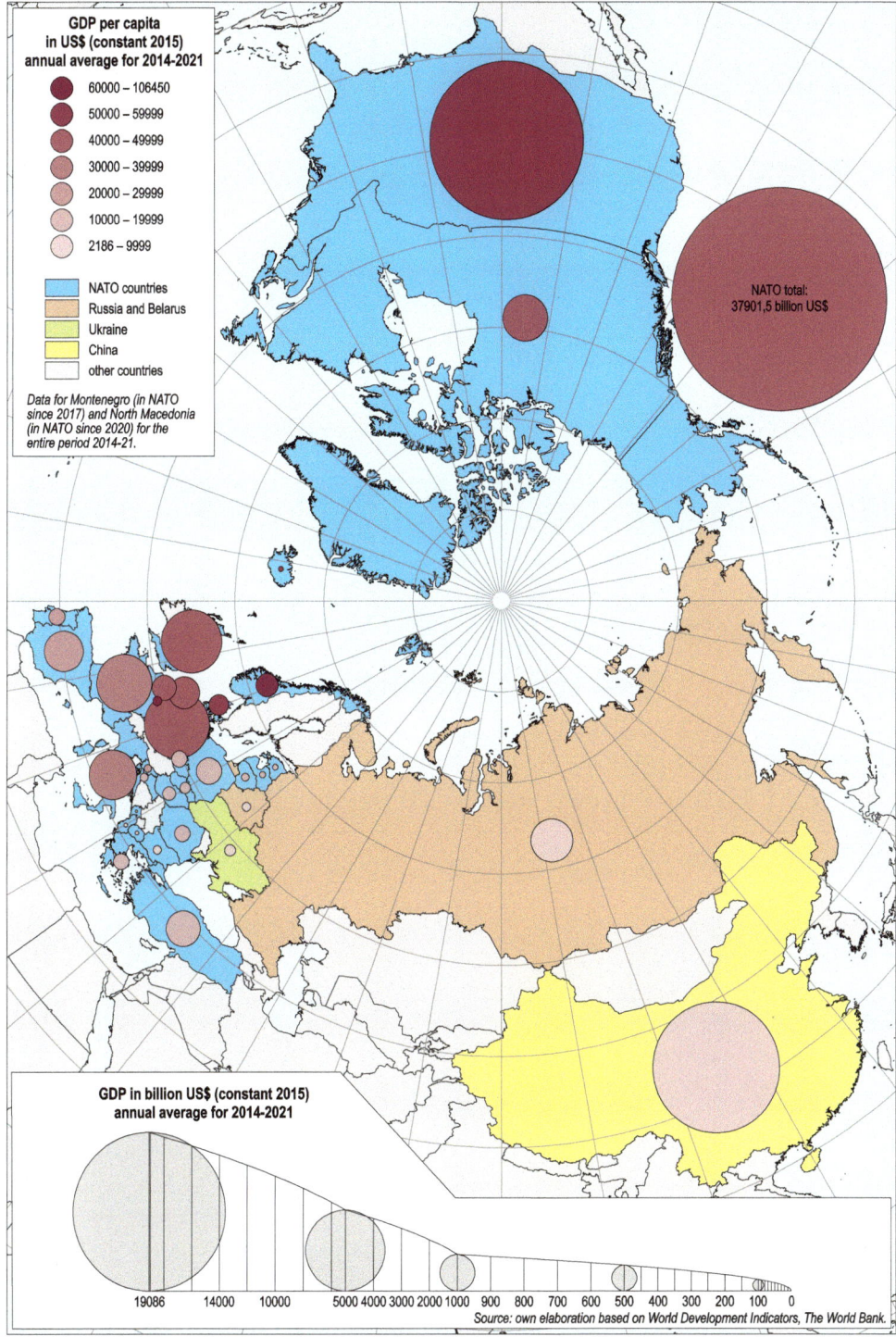

*Figure 10.25* GDP and per capita GDP: NATO and Ukraine vs Russia, Belarus, and the People's Republic of China: 2014–2021

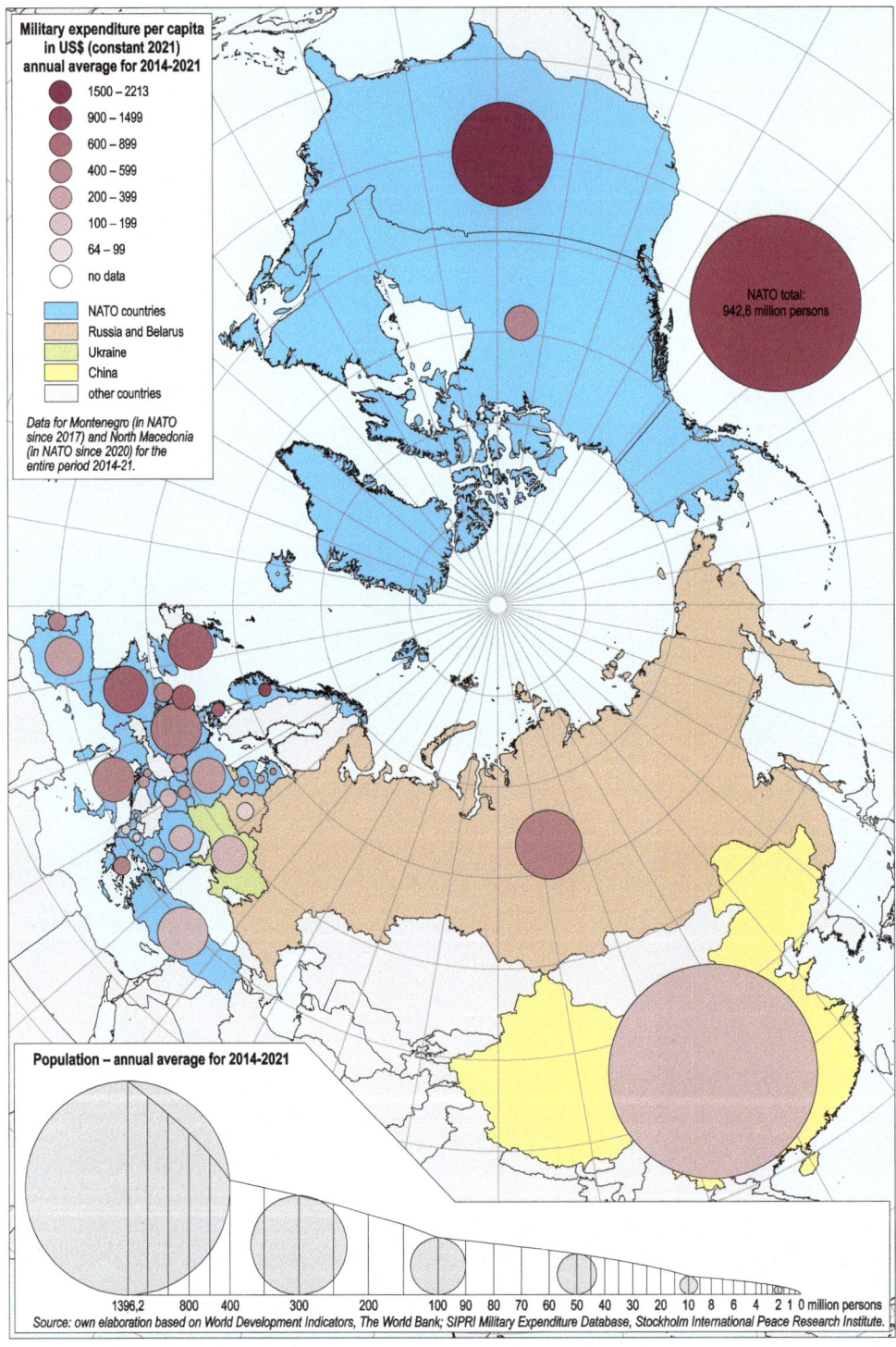

*Figure 10.26* Military expenditure per capita and population: NATO and Ukraine vs Russia, Belarus, and the People's Republic of China: 2014–2021

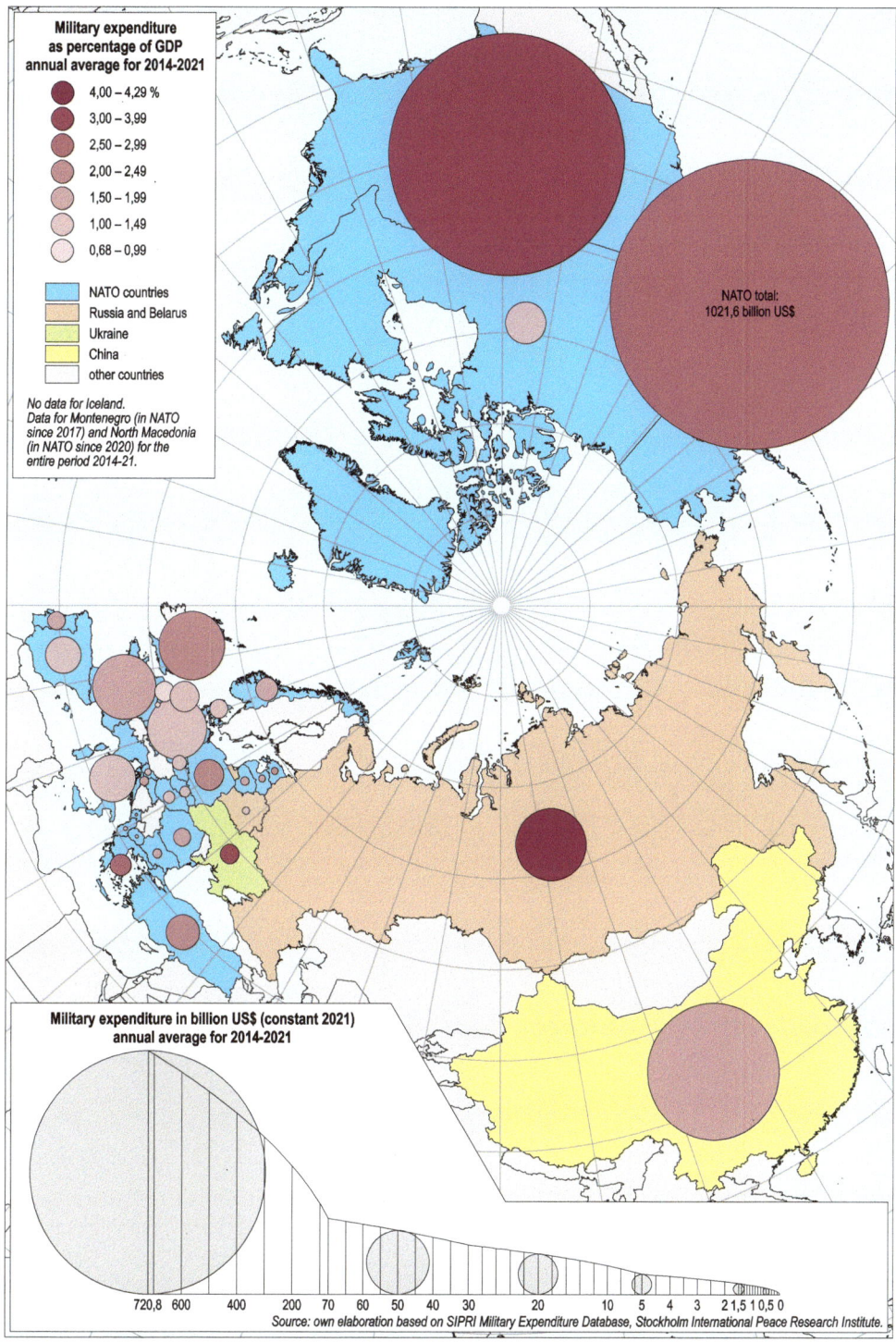

Military expenditure
as percentage of GDP
annual average for 2014-2021

- 4,00 – 4,29 %
- 3,00 – 3,99
- 2,50 – 2,99
- 2,00 – 2,49
- 1,50 – 1,99
- 1,00 – 1,49
- 0,68 – 0,99

NATO countries
Russia and Belarus
Ukraine
China
other countries

*No data for Iceland.
Data for Montenegro (in NATO
since 2017) and North Macedonia
(in NATO since 2020) for the
entire period 2014-21.*

NATO total:
1021,6 billion US$

Military expenditure in billion US$ (constant 2021)
annual average for 2014-2021

720,8 600   400    200  70 60 50  40  30    20    10   5   4   3  2 1,5 1 0,5 0

*Source: own elaboration based on SIPRI Military Expenditure Database, Stockholm International Peace Research Institute.*

*Figure 10.27* Military expenditure as a percentage of GDP and in absolute terms: NATO and Ukraine vs Russia, Belarus, and the People's Republic of China: 2014–2021

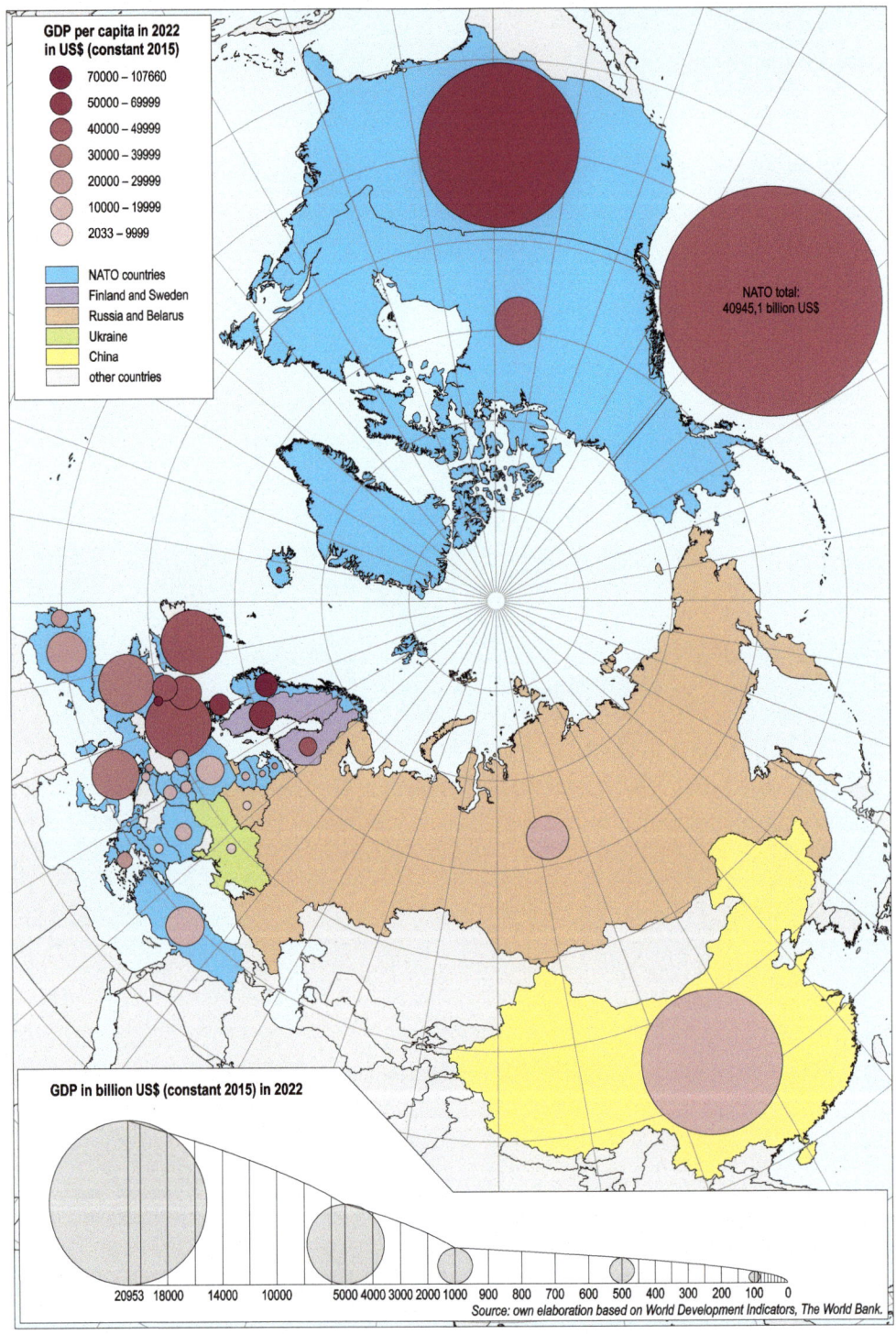

*Figure 10.28* GDP and per capita GDP: NATO and Ukraine vs Russia, Belarus, and the People's Republic of China: 2022

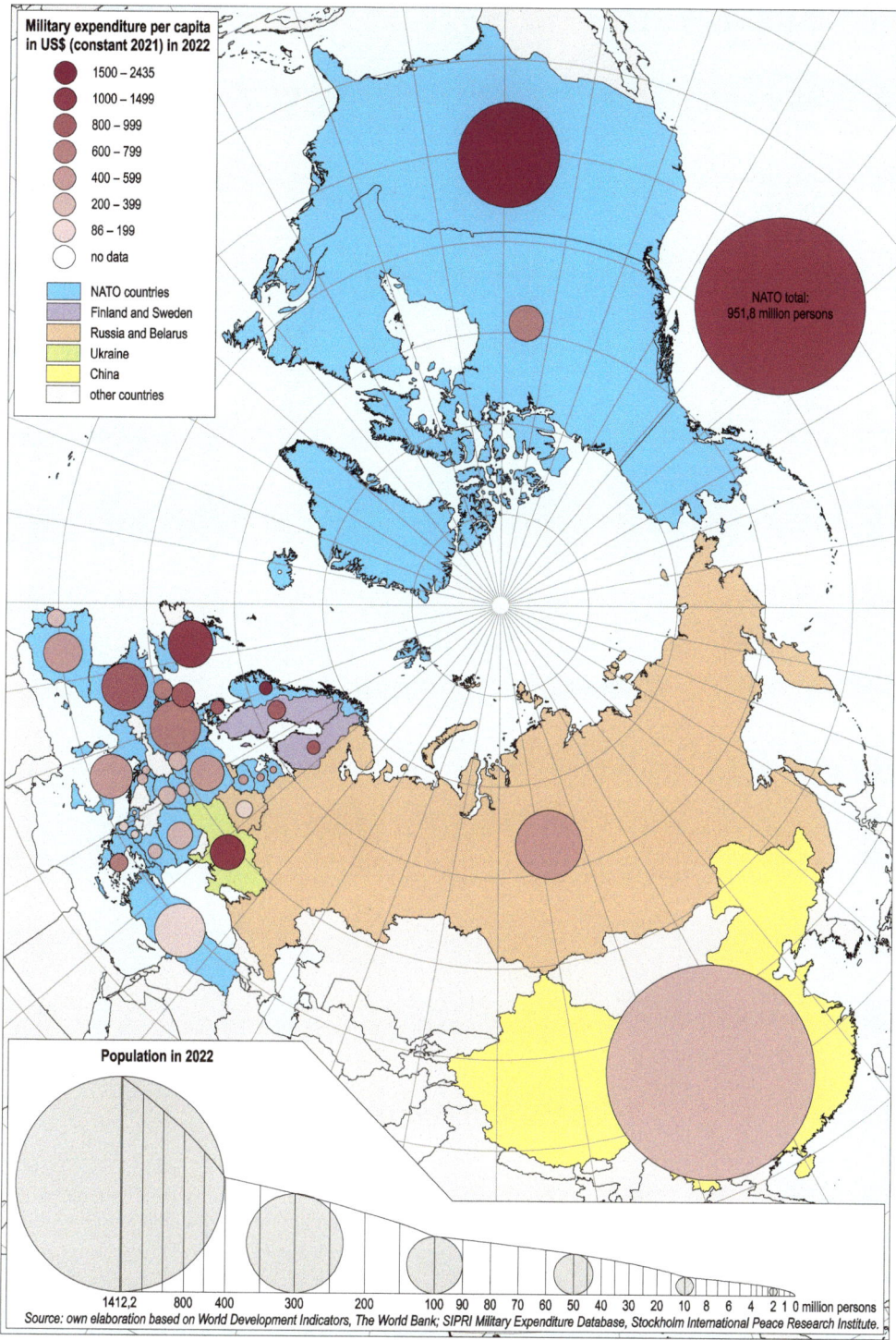

*Figure 10.29* Military expenditure per capita and population: NATO and Ukraine vs Russia, Belarus, and the People's Republic of China: 2022

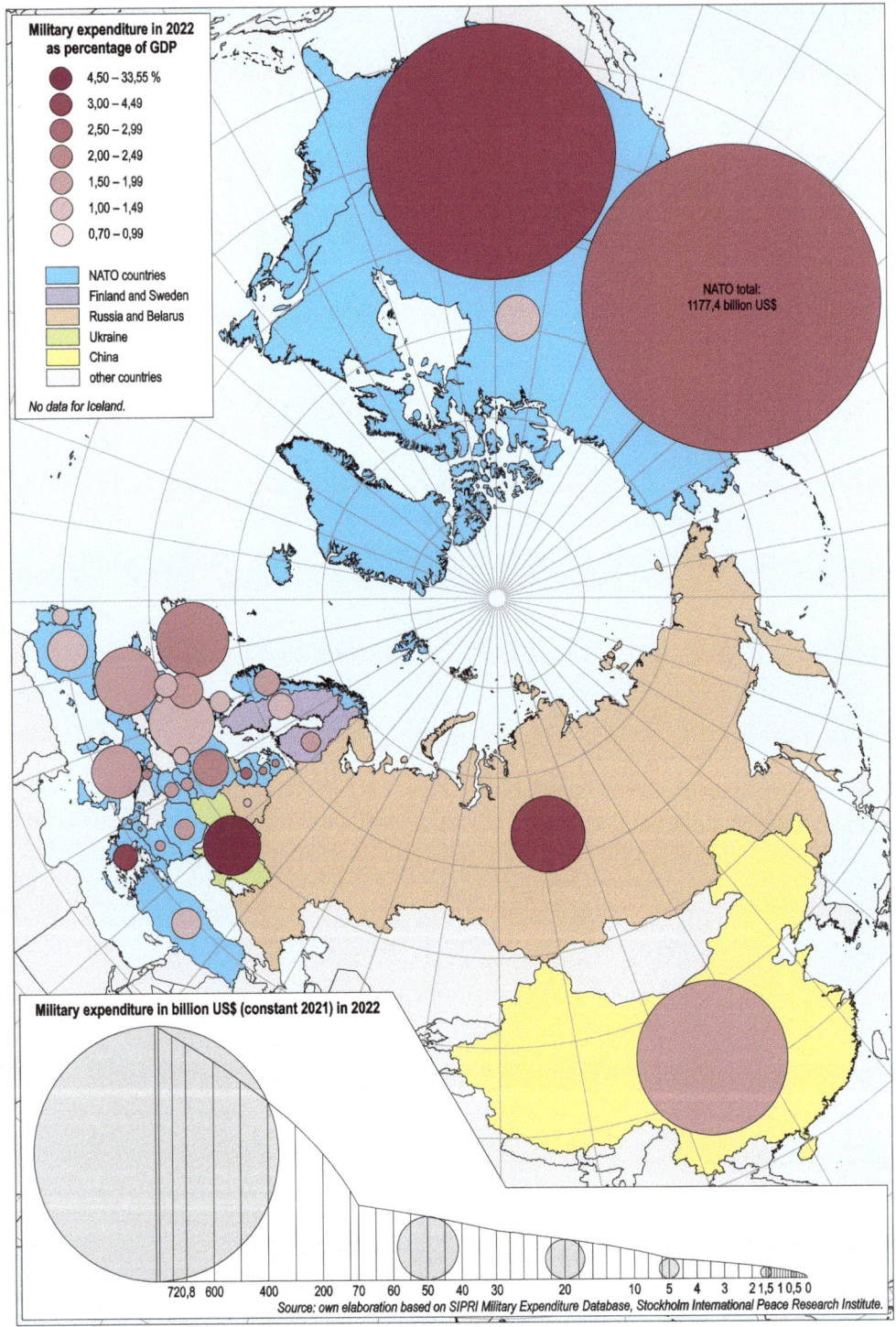

*Figure 10.30* Military expenditure as a percentage of GDP and in absolute terms: NATO and Ukraine vs Russia, Belarus, and the People's Republic of China: 2022

# 11 The international community voices

Most of the maps in this section refer to the UN mechanism to resolve Security Council deadlocks arising from permanent members exercising their veto power. In these situations, the UN General Assembly is empowered to take action on important world affairs. These maps can therefore be used to follow the position of almost the entire international community (identified here with UN member states) in the face of significant international political issues since the 1950s. The maps in this series reveal informal alliances and opposing opinions, attitudes, and interests. The palette of decisions analysed includes abstentions and absenteeism, as well as votes for and against. These maps correlate votes with the population and wealth of the countries casting them, thereby allowing for a broader and deeper insight into the opinions of the international community on major international issues since 1945.

This group of cartographic compilations additionally includes three maps showing politically motivated boycotts of important global sports events such as the Olympic Games. These maps also reveal various conditions, arrangements, and international ties.

Last but not least this group is supplemented by two maps, drawn in a similar style, that show the position of selected societies towards the war in Ukraine via two questions – one general, the other specific. Contrary to appearances, these actually concern the same issue. The answers given not only reveal global differences in attitude to the war in Ukraine but also different approaches to sovereignty and aggression. These maps additionally show the coherence of views or the hypocrisy of the societies under investigation. Above all, however, they may well point up the role of geography (the relevance of proximity and neighbourhood), history, and current politics in shaping political attitudes. This makes their interpretation very complex and challenging.

DOI: 10.4324/9781003406440-11

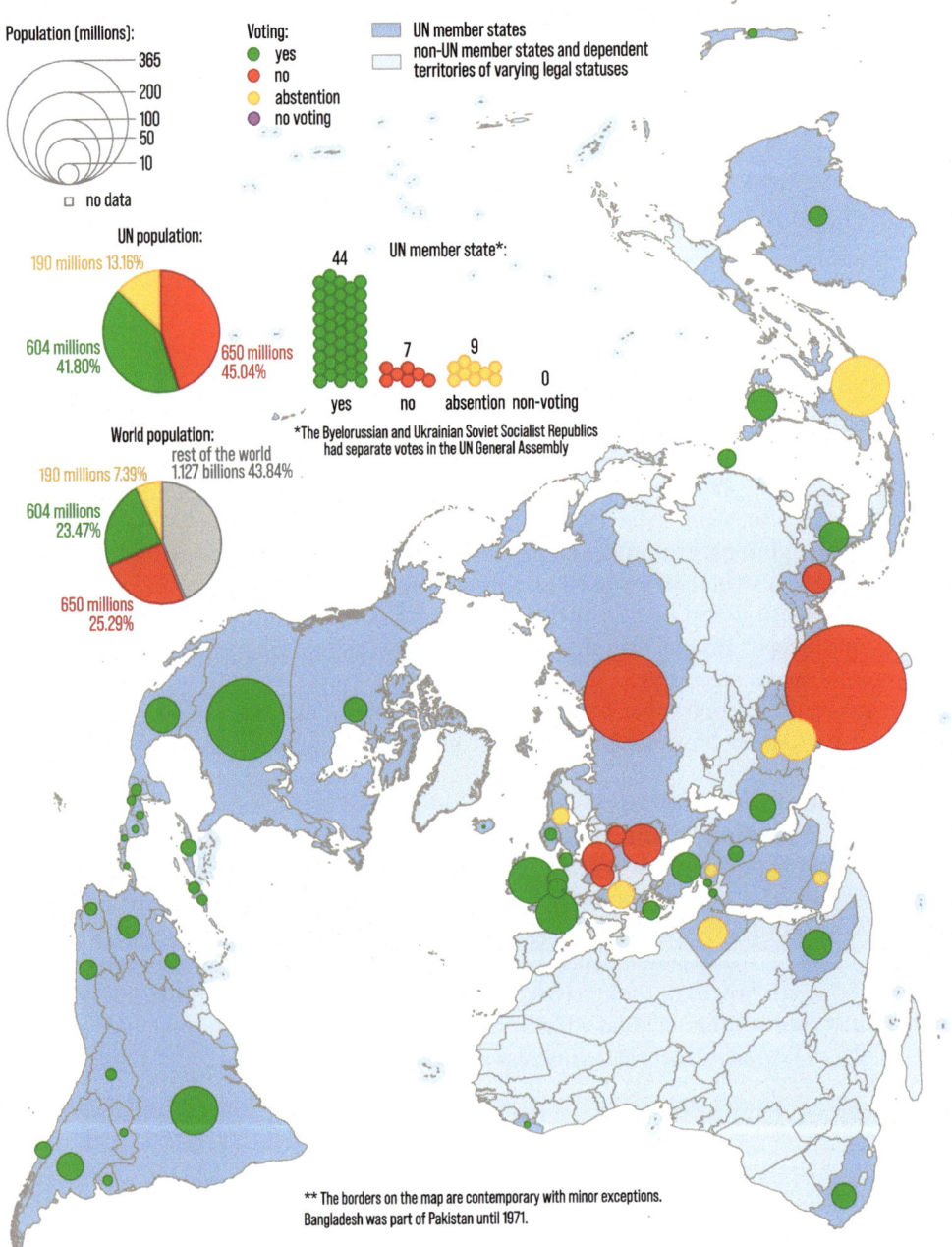

*Figure 11.1* 1951: UN resolution regarding the intervention of the People's Republic of China in Korea [A/RES/498(V)]. The UN General Assembly deems the PRC to be engaged in aggression and demands the withdrawal of Chinese troops and citizens from Korea. Voting according to UN and world population.

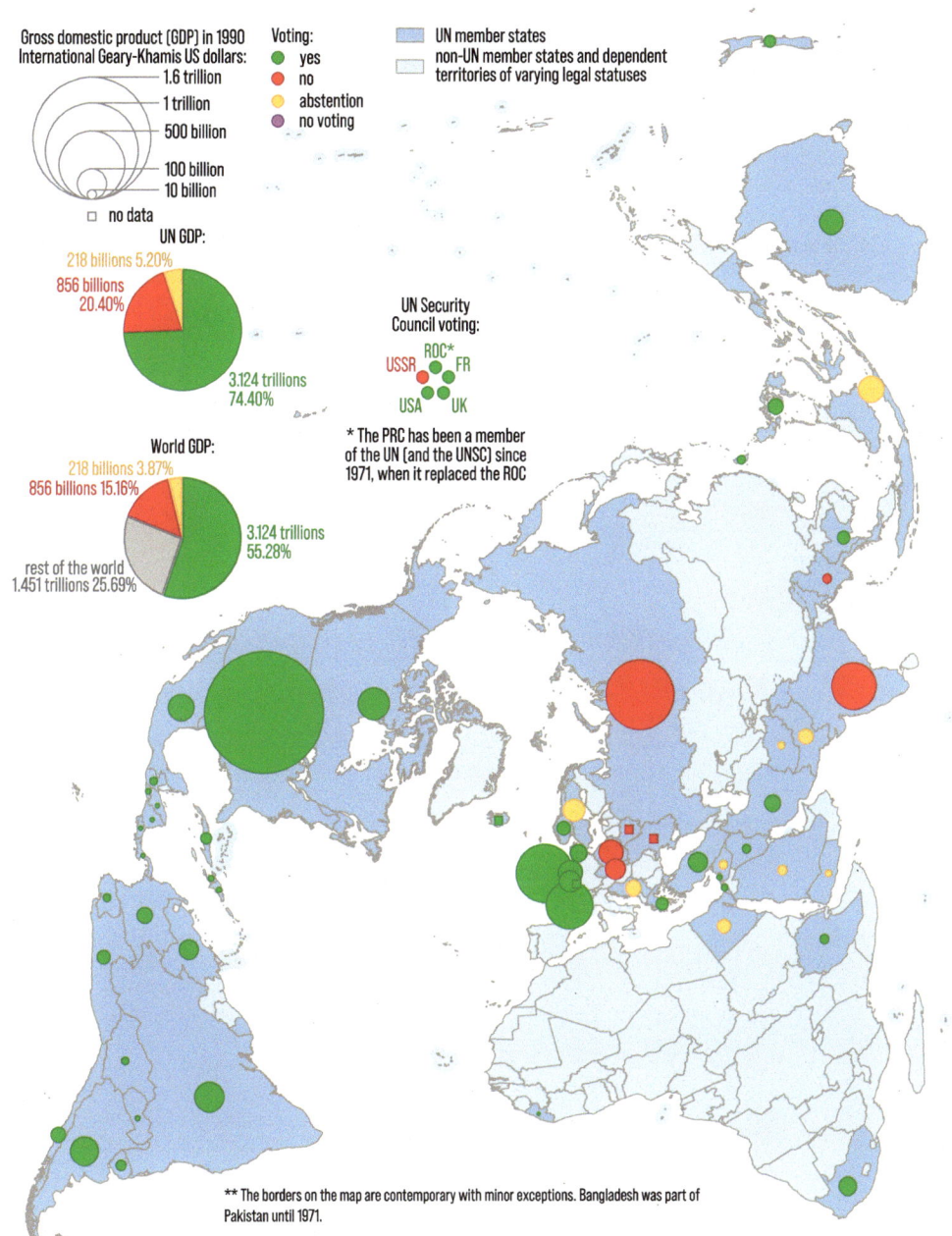

Gross domestic product (GDP) in 1990 International Geary-Khamis US dollars:
- 1.6 trillion
- 1 trillion
- 500 billion
- 100 billion
- 10 billion
- □ no data

Voting:
- ● yes
- ● no
- ● abstention
- ● no voting

■ UN member states
□ non-UN member states and dependent territories of varying legal statuses

UN GDP:
- 218 billions 5.20%
- 856 billions 20.40%
- 3.124 trillions 74.40%

World GDP:
- 218 billions 3.87%
- 856 billions 15.16%
- rest of the world 1.451 trillions 25.69%
- 3.124 trillions 55.28%

UN Security Council voting:
USSR ROC* FR
USA UK

* The PRC has been a member of the UN (and the UNSC) since 1971, when it replaced the ROC

** The borders on the map are contemporary with minor exceptions. Bangladesh was part of Pakistan until 1971.

*Source:* Own elaboration based on United Nations digital library (https://digitallibrary.un.org); A. Maddison, Maddison Database 2010 (www.rug.nl/ggdc/)

*Figure 11.2* 1951: UN resolution regarding the intervention of the People's Republic of China in Korea [A/RES/498(V)]. The UN General Assembly deems the PRC to be engaged in aggression and demands the withdrawal of Chinese troops and citizens from Korea. Voting according to UN and world GDP.

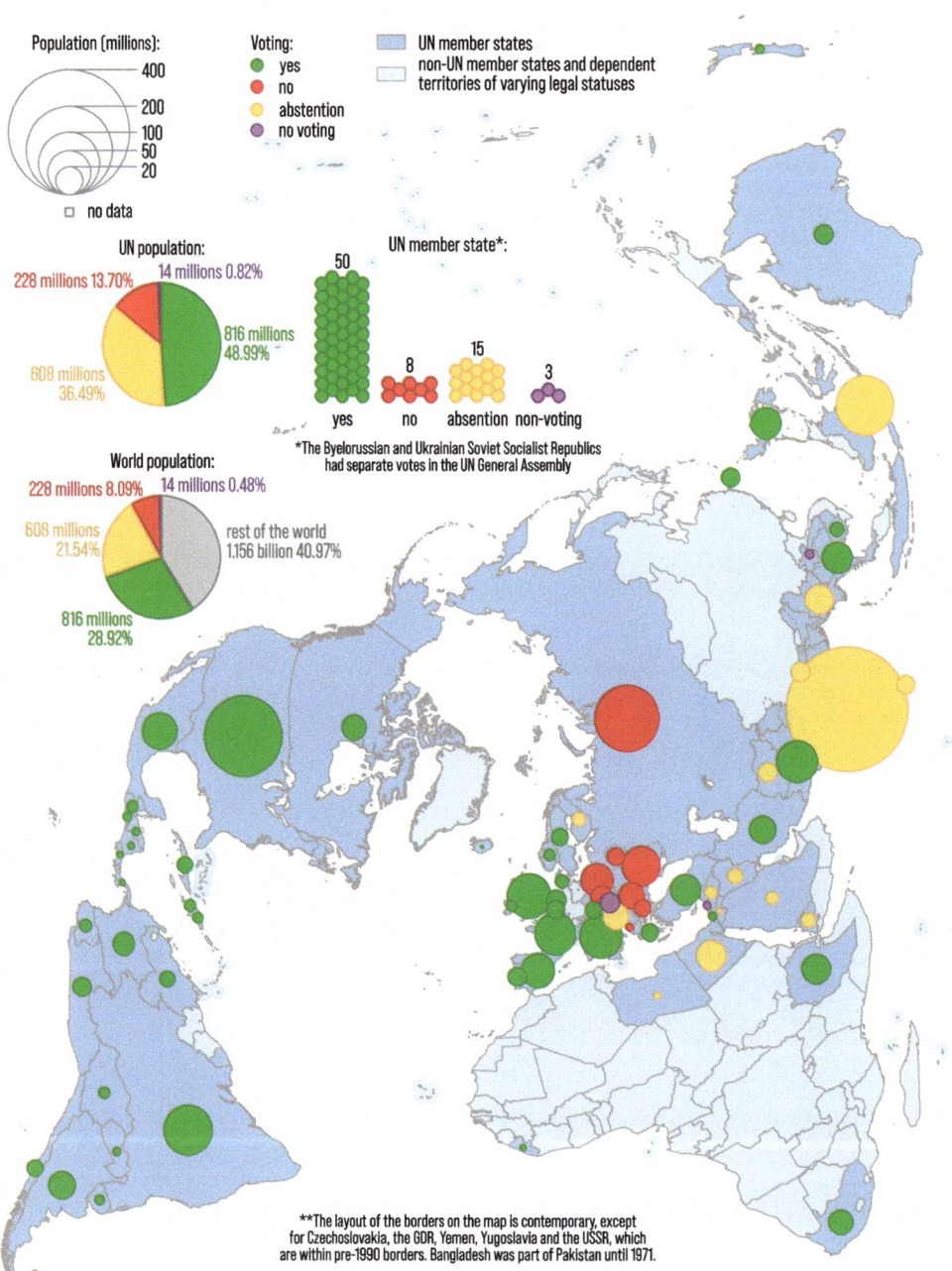

Population (millions):
400
200
100
50
20
□ no data

Voting:
● yes
● no
● abstention
● no voting

UN member states
non-UN member states and dependent territories of varying legal statuses

UN population:
228 millions 13.70%
608 millions 36.49%
816 millions 48.99%
14 millions 0.82%

UN member state*:
50 yes
8 no
15 abstention
3 non-voting
*The Byelorussian and Ukrainian Soviet Socialist Republics had separate votes in the UN General Assembly

World population:
228 millions 8.09%
608 millions 21.54%
816 millions 28.92%
14 millions 0.48%
rest of the world 1.156 billion 40.97%

**The layout of the borders on the map is contemporary, except for Czechoslovakia, the GDR, Yemen, Yugoslavia and the USSR, which are within pre-1990 borders. Bangladesh was part of Pakistan until 1971.

*Source:* Own elaboration based on United Nations digital library (https://digitallibrary.un.org); A. Maddison, Maddison Database 2010 (www.rug.nl/ggdc/)

*Figure 11.3* 1956: UN resolution regarding Soviet intervention in Hungary (1004/ES-11). The UN General Assembly calls on the USSR to cease its intervention in Hungary's internal affairs and to withdraw its army. Voting according to UN and world population.

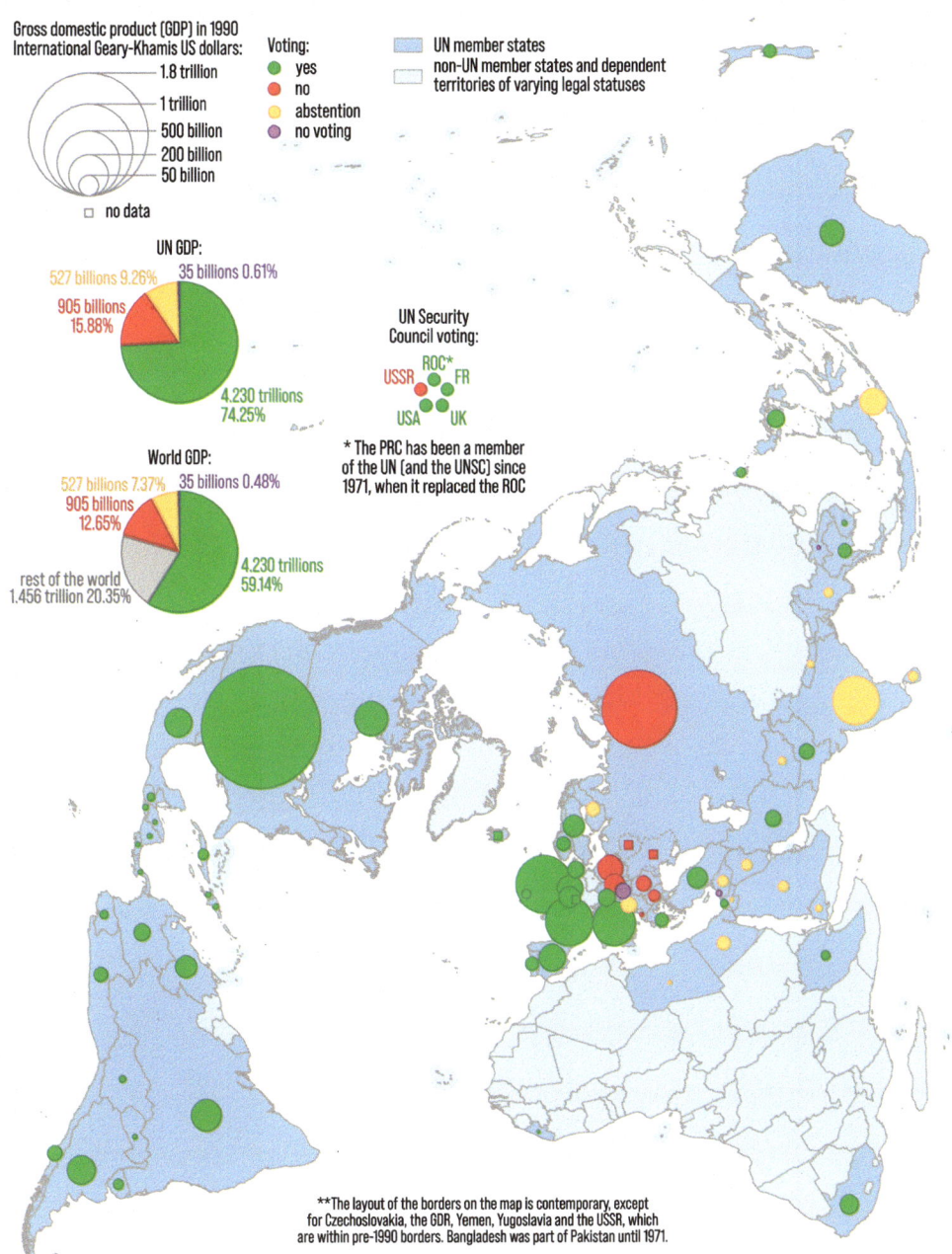

*Figure 11.4* 1956: UN resolution regarding Soviet intervention in Hungary (1004/ES-11). The UN General Assembly calls on the USSR to cease its intervention in Hungary's internal affairs and to withdraw its army. Voting according to UN and world GDP.

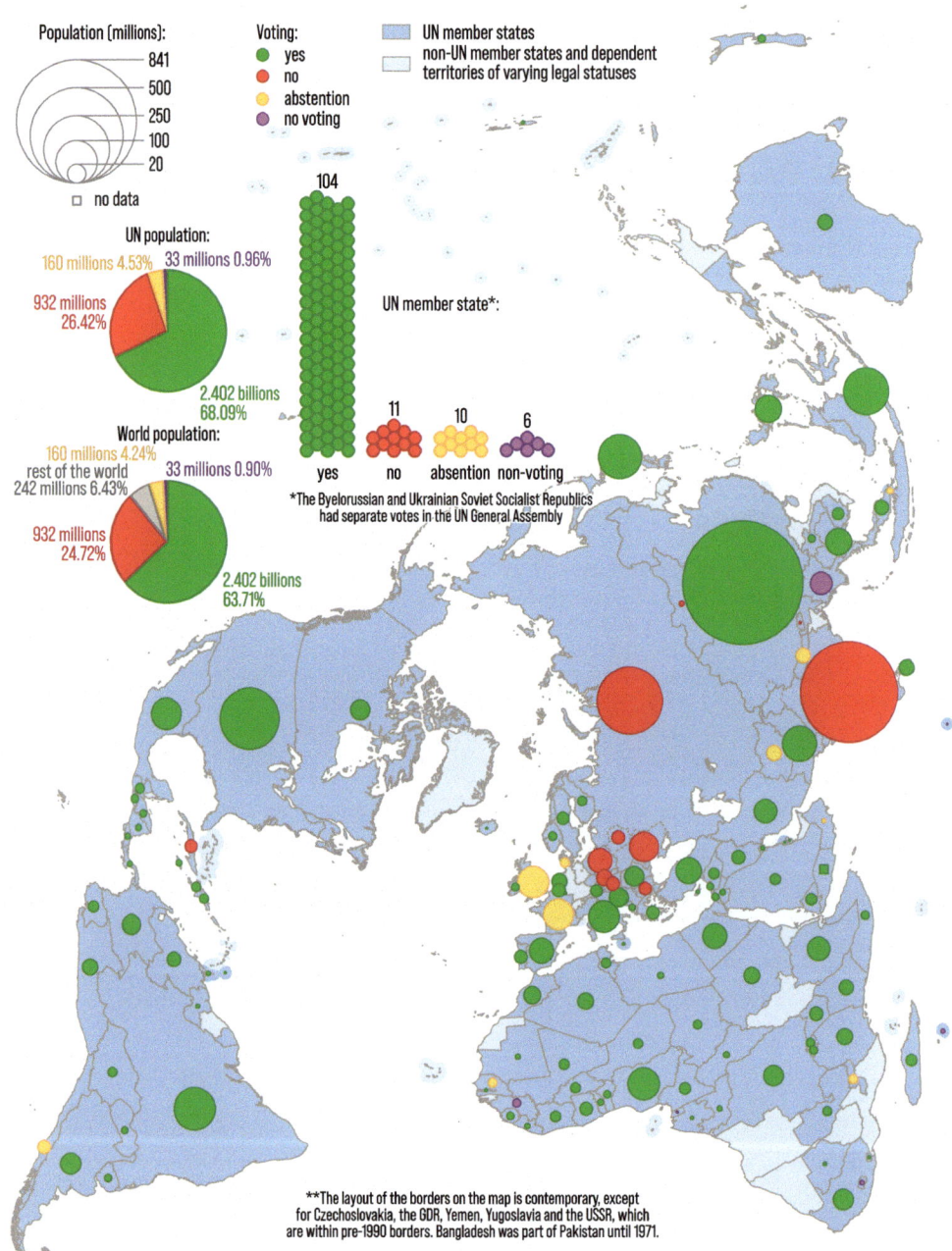

*Figure 11.5* 1971: UN resolution regarding the Indo-Pakistani war over Bangladesh [A/RES/2793(XXVI)]. The UN General Assembly calls for an immediate cease-fire and for both countries to withdraw their troops. Voting according to UN and world population.

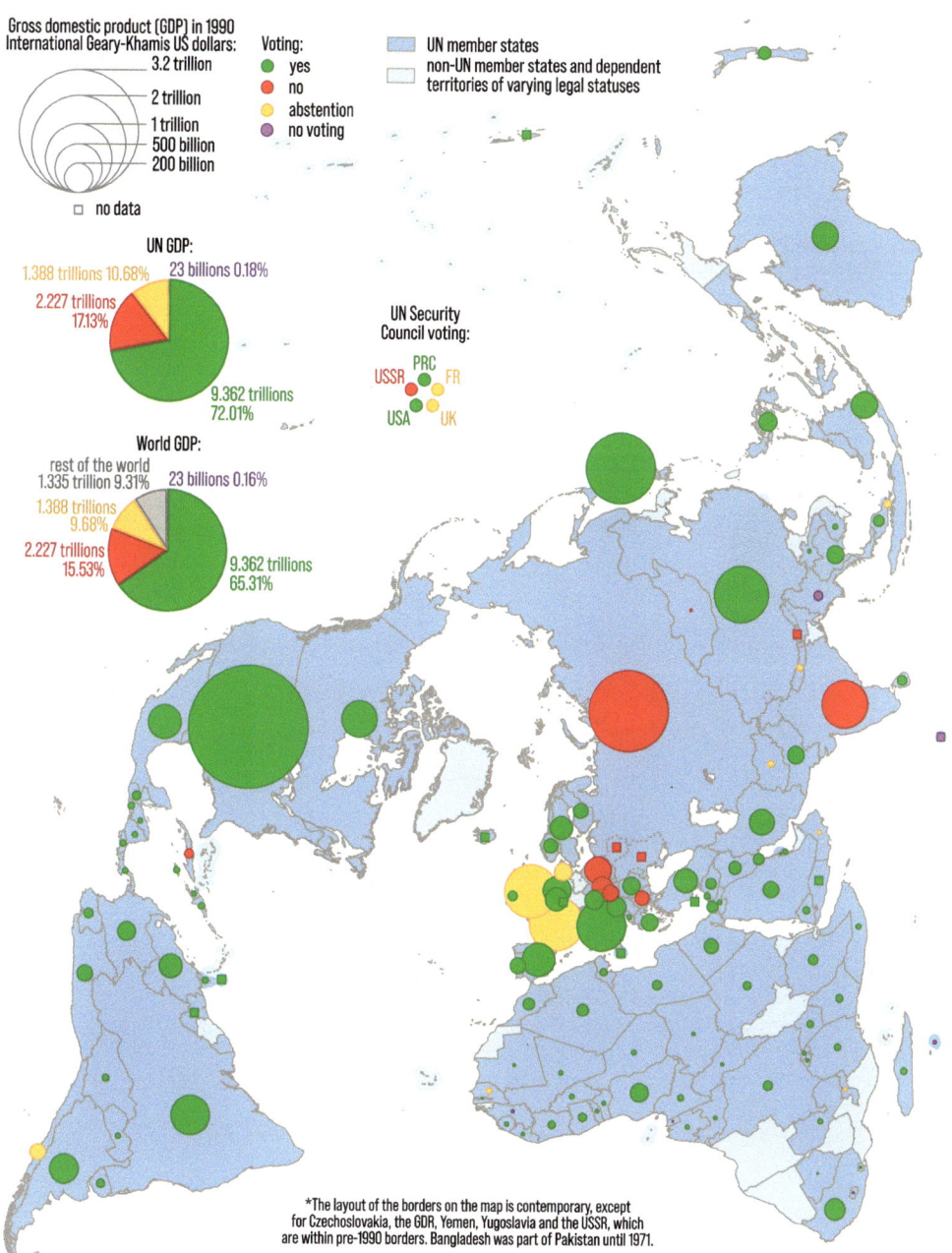

*Figure 11.6* 1971: UN resolution regarding the Indo-Pakistani war over Bangladesh [A/RES/2793(XXVI)]. The UN General Assembly calls for an immediate cease-fire and for both countries to withdraw their troops. Voting according to UN and world GDP.

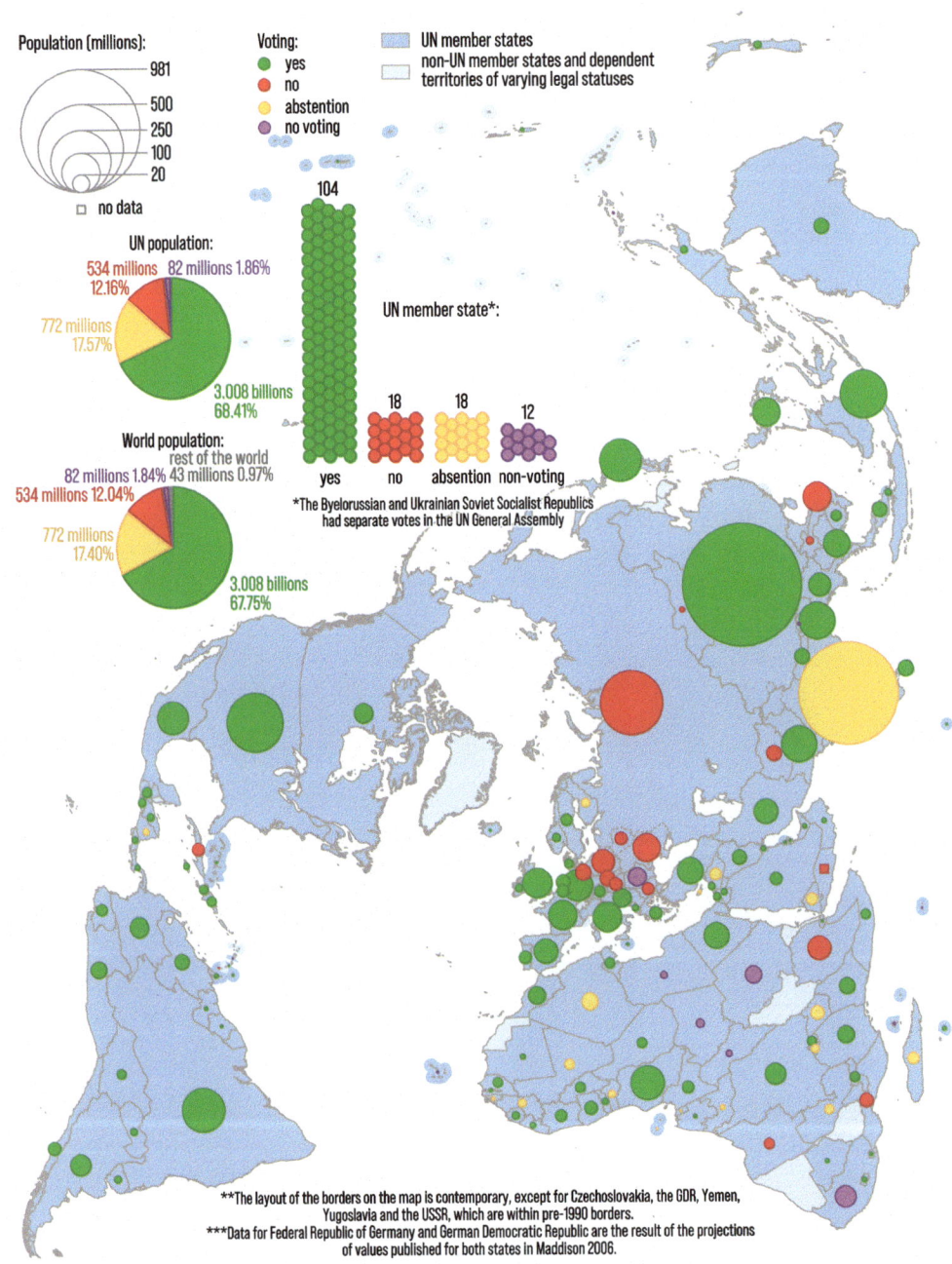

*Figure 11.7* 1980: UN resolution regarding the Soviet aggression against Afghanistan (ES-6/2). The UN General Assembly calls for withdrawal of Soviet troops from Afghanistan. Voting according to UN and world population.

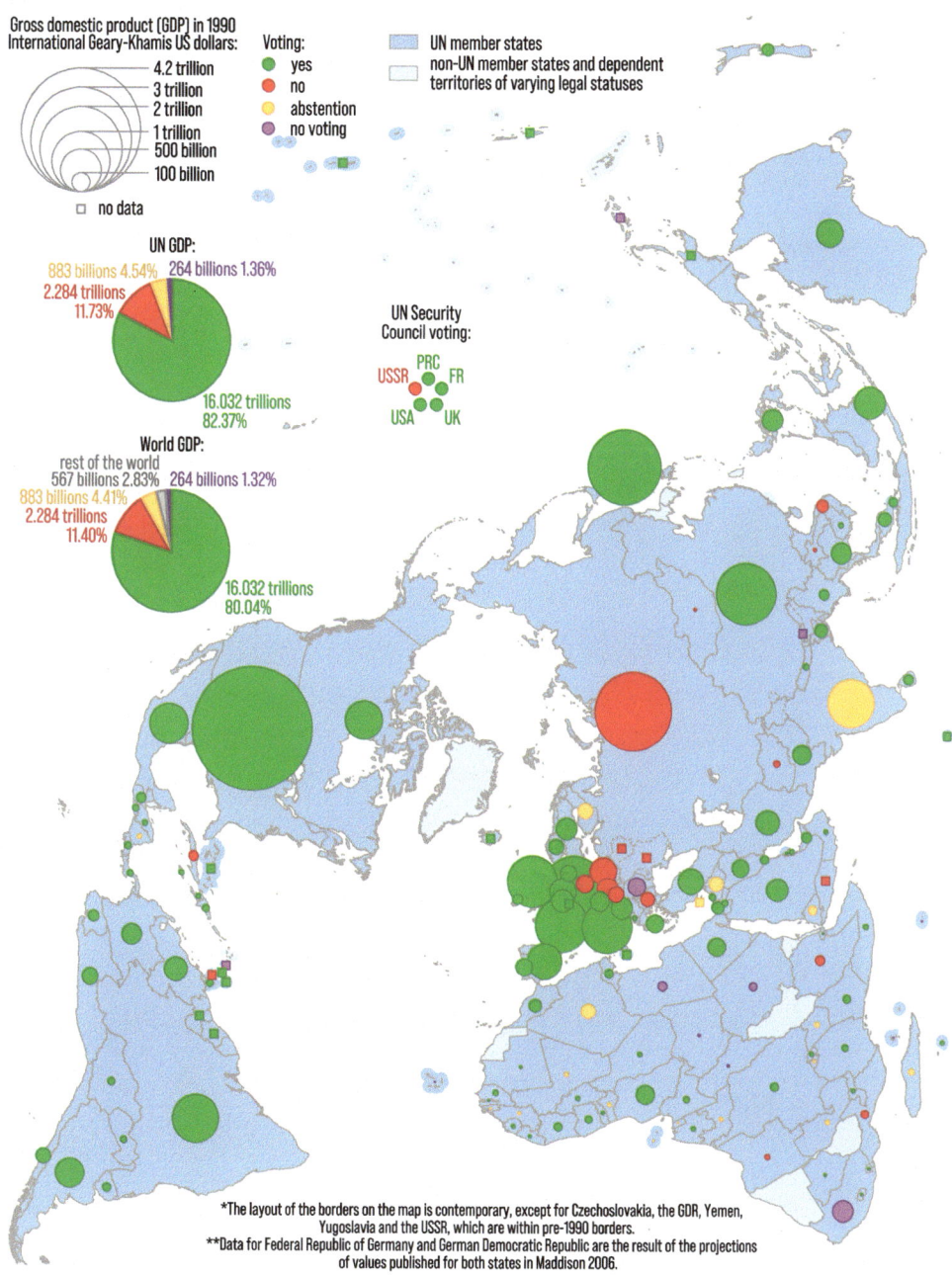

*Figure 11.8* 1980: UN resolution regarding the Soviet aggression against Afghanistan (ES-6/2). The UN General Assembly calls for withdrawal of Soviet troops from Afghanistan. Voting according to UN and world GDP.

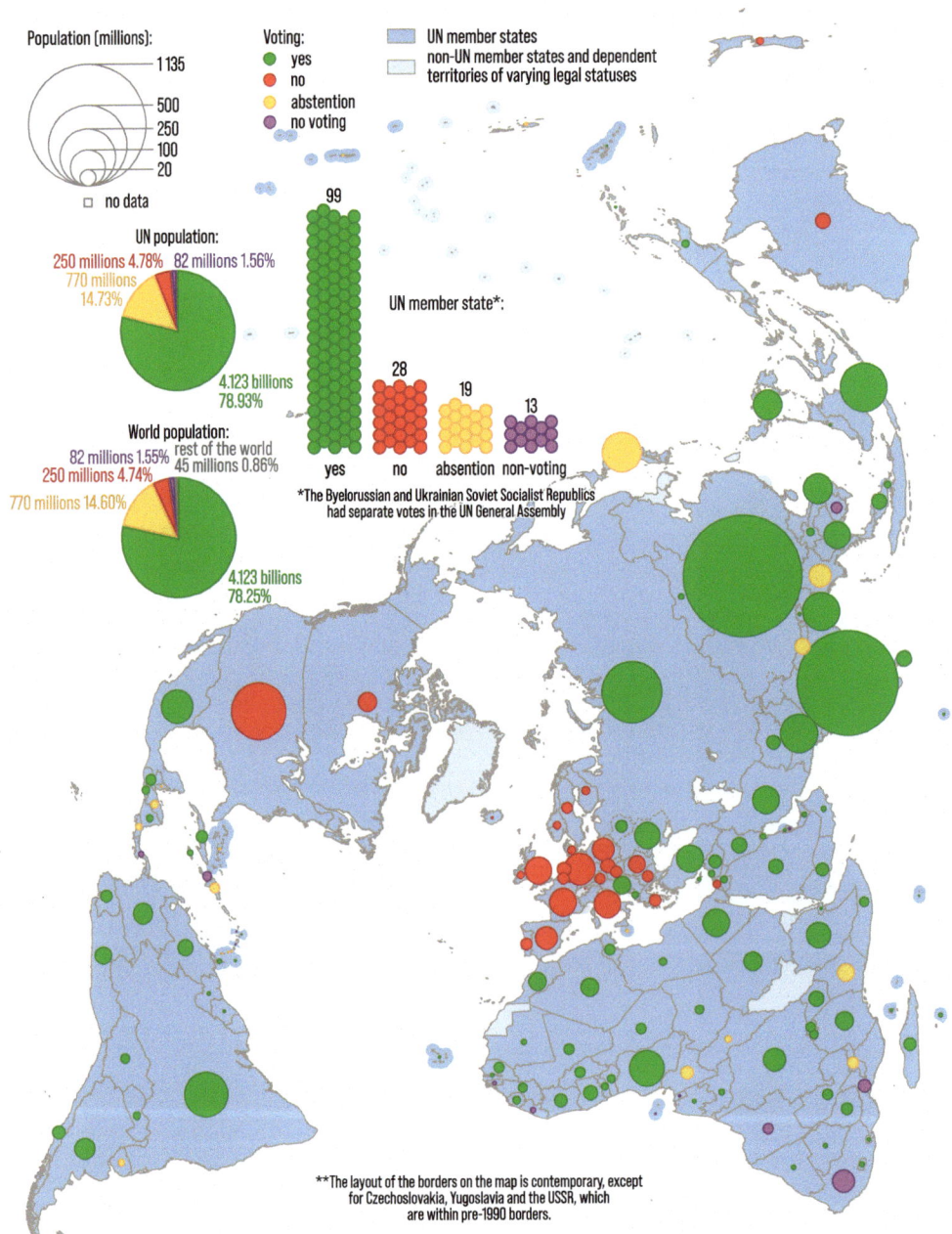

*Figure 11.9* 1990: UN resolution regarding apartheid and Israel (A/RES/45/176D). The UN General Assembly condemns Israel's military and nuclear cooperation with South Africa. Voting according to UN and world population.

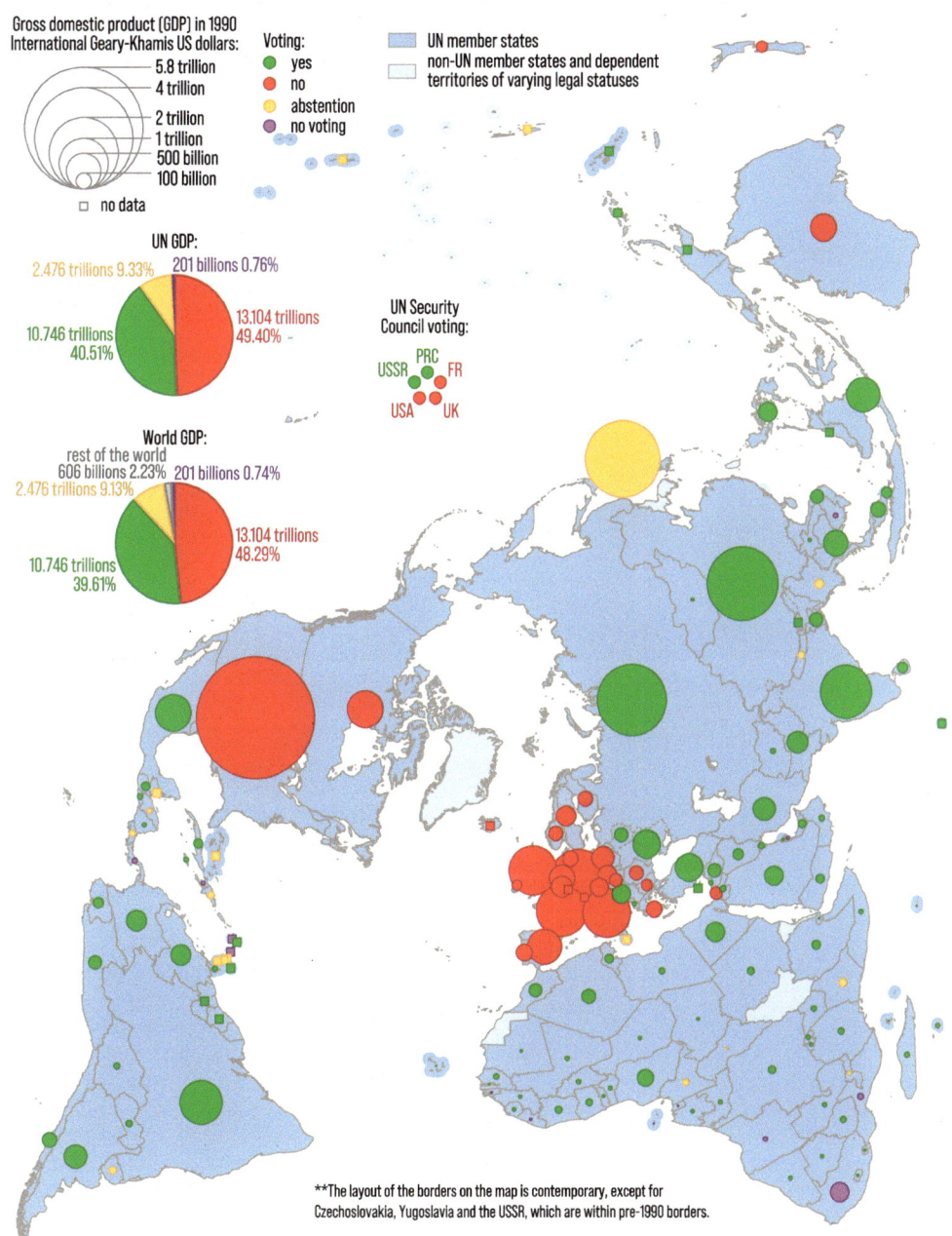

Gross domestic product (GDP) in 1990
International Geary-Khamis US dollars:

5.8 trillion
4 trillion
2 trillion
1 trillion
500 billion
100 billion

□ no data

Voting:
● yes
● no
● abstention
● no voting

UN member states
non-UN member states and dependent territories of varying legal statuses

UN GDP:
201 billions 0.76%
2.476 trillions 9.33%
13.104 trillions 49.40%
10.746 trillions 40.51%

UN Security Council voting:
USSR PRC FR
USA UK

World GDP:
rest of the world
606 billions 2.23%   201 billions 0.74%
2.476 trillions 9.13%
13.104 trillions 48.29%
10.746 trillions 39.61%

**The layout of the borders on the map is contemporary, except for Czechoslovakia, Yugoslavia and the USSR, which are within pre-1990 borders.

*Source:* Own elaboration based on United Nations digital library (https://digitallibrary.un.org); A. Maddison, Maddison Database 2010 (www.rug.nl/ggdc/)

*Figure 11.10* 1990: UN resolution regarding apartheid and Israel (A/RES/45/176D). The UN General Assembly condemns Israel's military and nuclear cooperation with South Africa. Voting according to UN and world GDP.

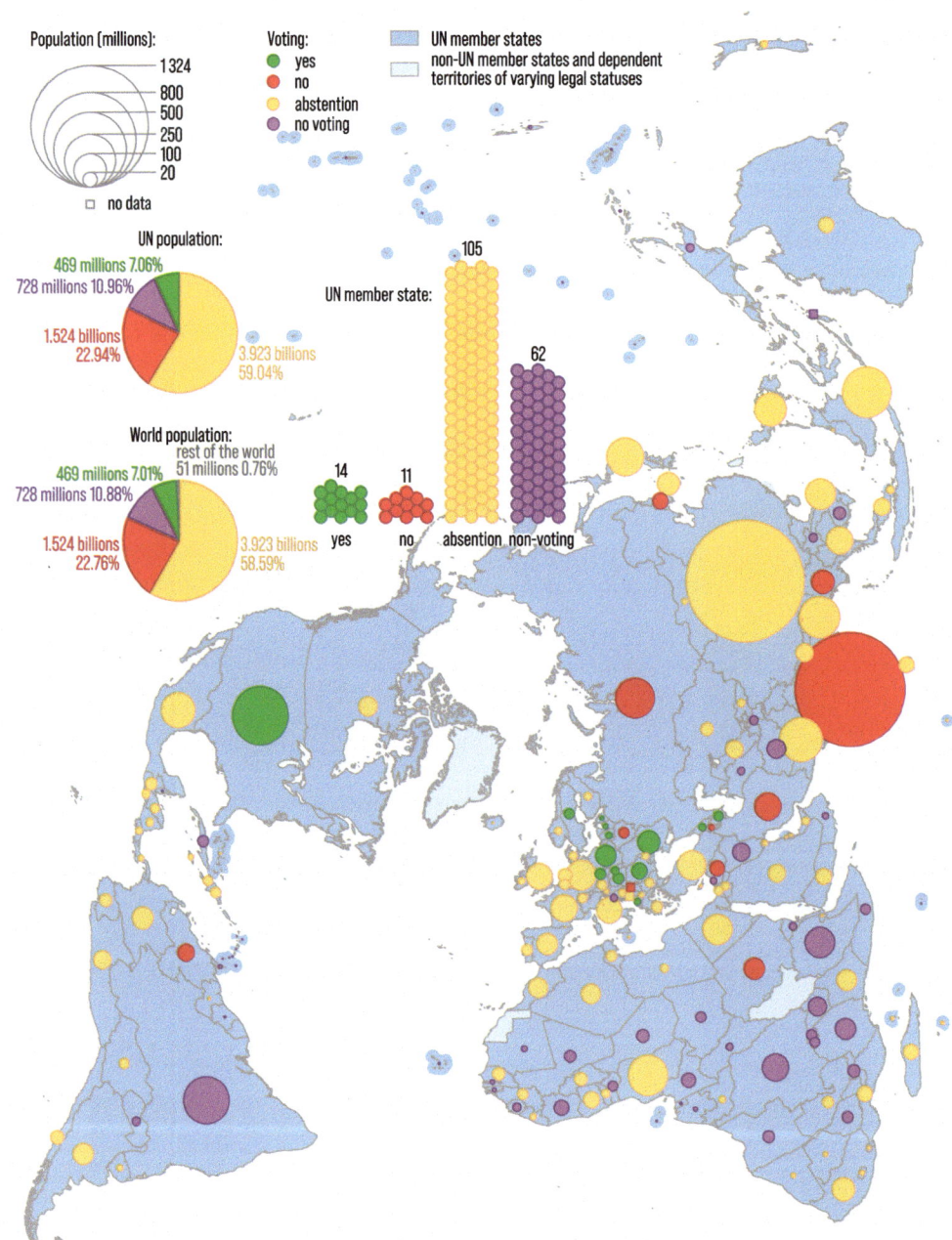

*Figure 11.11* 2008: UN resolution regarding the conflict in Georgia (A/RES/62/249). The UN General Assembly recognises the right of return of all refugees and internally displaced persons and their descendants, regardless of ethnicity, to Abkhazia, Georgia. Voting according to UN and world population.

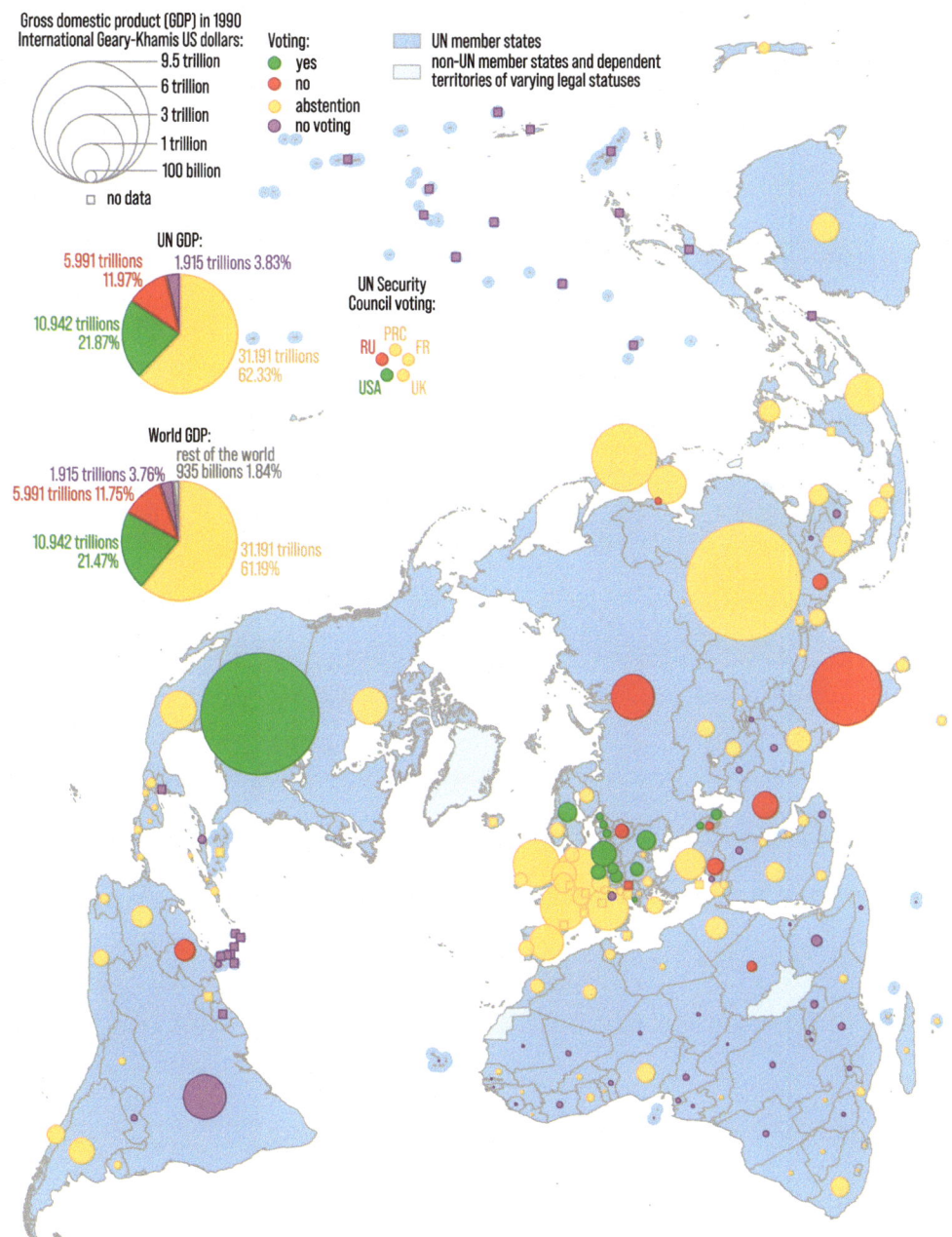

*Figure 11.12*  2008: UN resolution regarding the conflict in Georgia (A/RES/62/249). The UN General Assembly recognises the right of return of all refugees and internally displaced persons and their descendants, regardless of ethnicity, to Abkhazia, Georgia. Voting according to UN and world GDP.

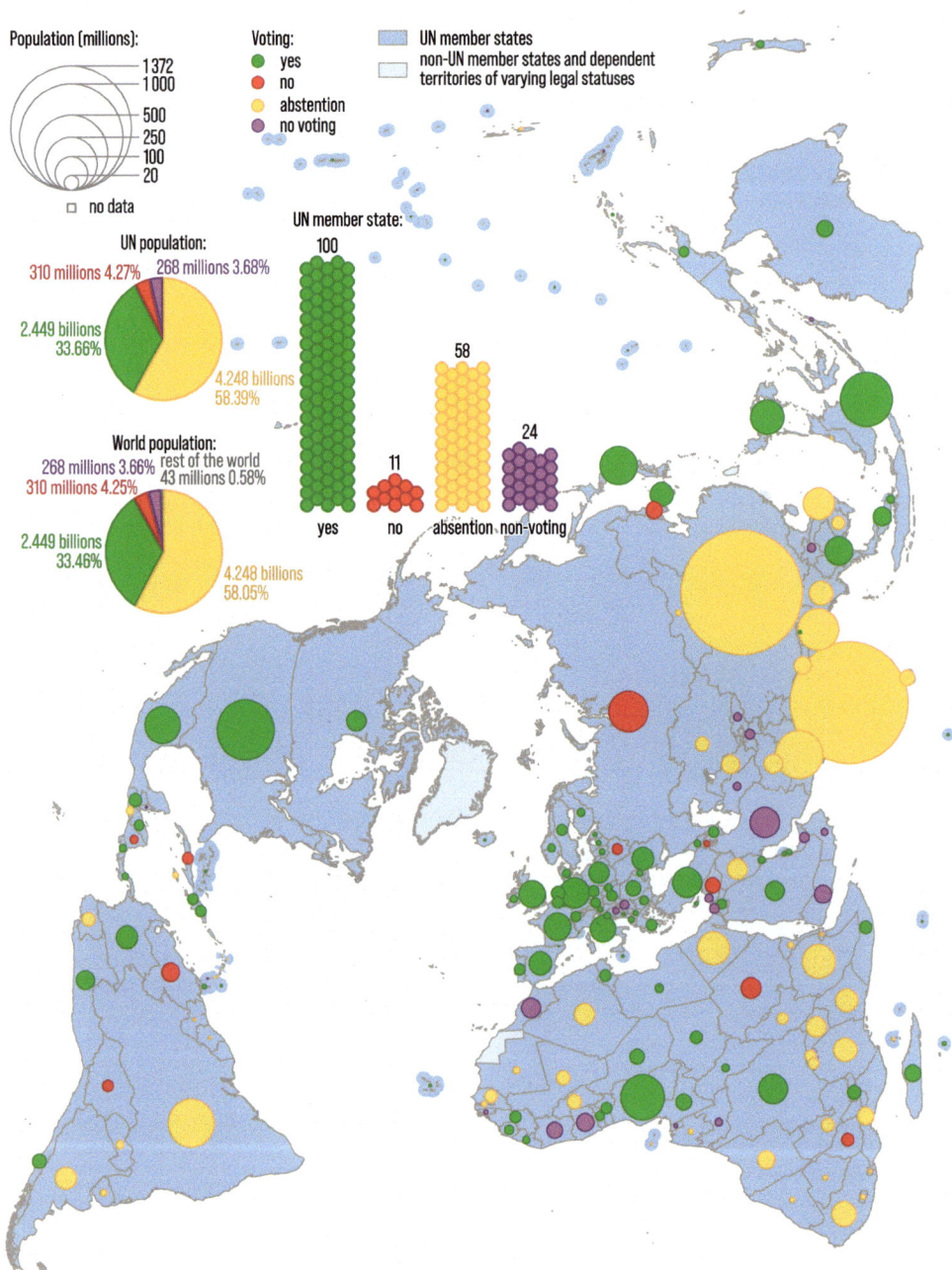

Source: Own elaboration based on United Nations digital library (https://digitallibrary.un.org); The World Bank data (https://databank.worldbank.org)

*Figure 11.13* 2014: UN resolution regarding the Russo-Ukrainian War (first episode) (A/RES/68/262). The UN General Assembly affirms Ukraine's territorial integrity within its internationally recognised borders and calls for them to be respected. Voting according to UN and world population.

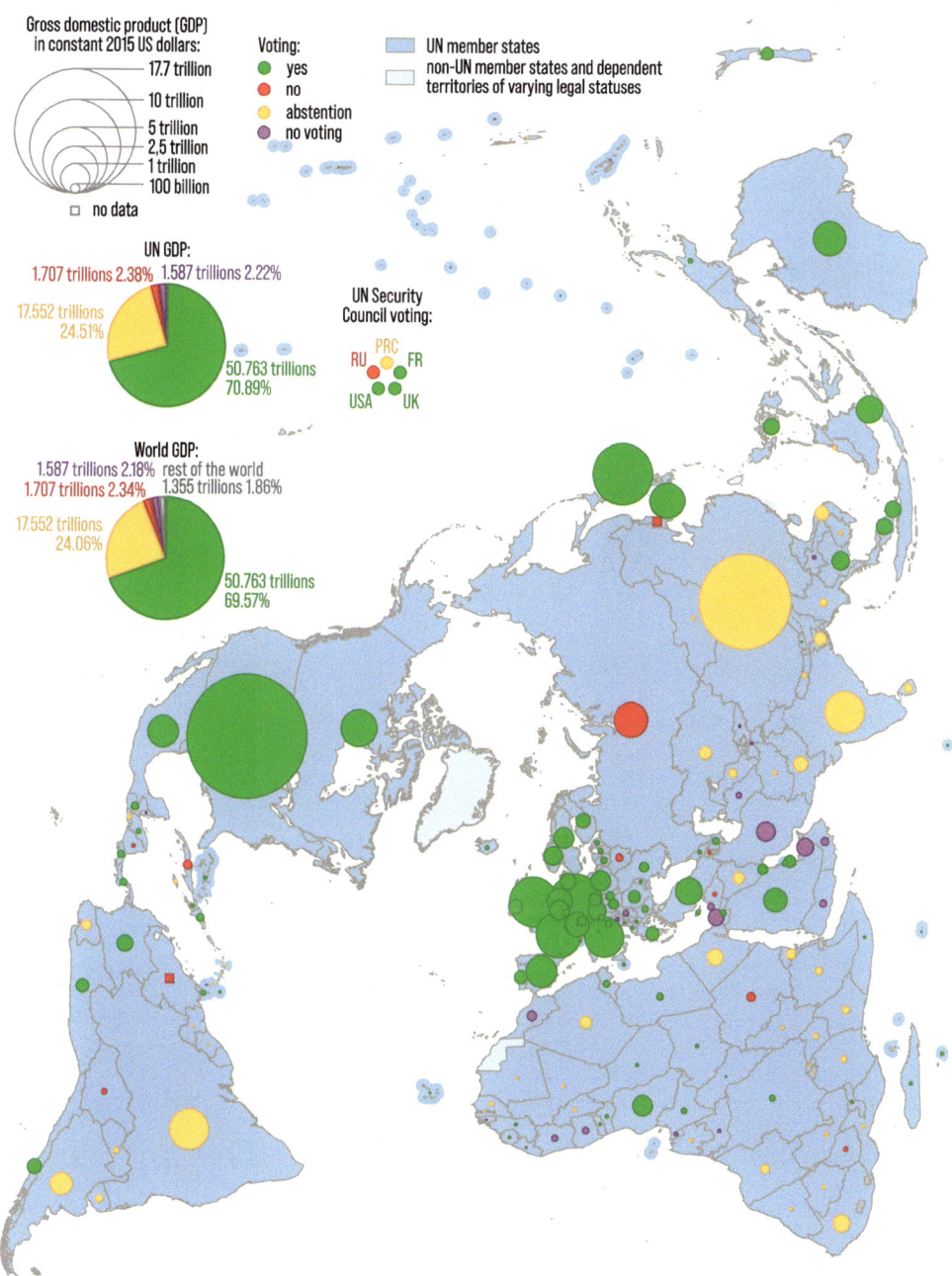

Gross domestic product (GDP)
in constant 2015 US dollars:
17.7 trillion
10 trillion
5 trillion
2,5 trillion
1 trillion
100 billion
□ no data

Voting:
● yes
● no
● abstention
● no voting

■ UN member states
□ non-UN member states and dependent
territories of varying legal statuses

UN GDP:
1.707 trillions 2.38%  1.587 trillions 2.22%
17.552 trillions 24.51%
50.763 trillions 70.89%

UN Security
Council voting:
RU  PRC  FR
USA  UK

World GDP:
1.587 trillions 2.18%  rest of the world
1.707 trillions 2.34%  1.355 trillions 1.86%
17.552 trillions 24.06%
50.763 trillions 69.57%

Source: Own elaboration based on United Nations digital library (https://digitallibrary.un.org); The World Bank data (https://databank.worldbank.org)

*Figure 11.14* 2014: UN resolution regarding the Russo-Ukrainian War (first episode) (A/
RES/68/262). The UN General Assembly affirms Ukraine's territorial integrity within
its internationally recognised borders and calls for them to be respected. Voting
according to UN and world GDP.

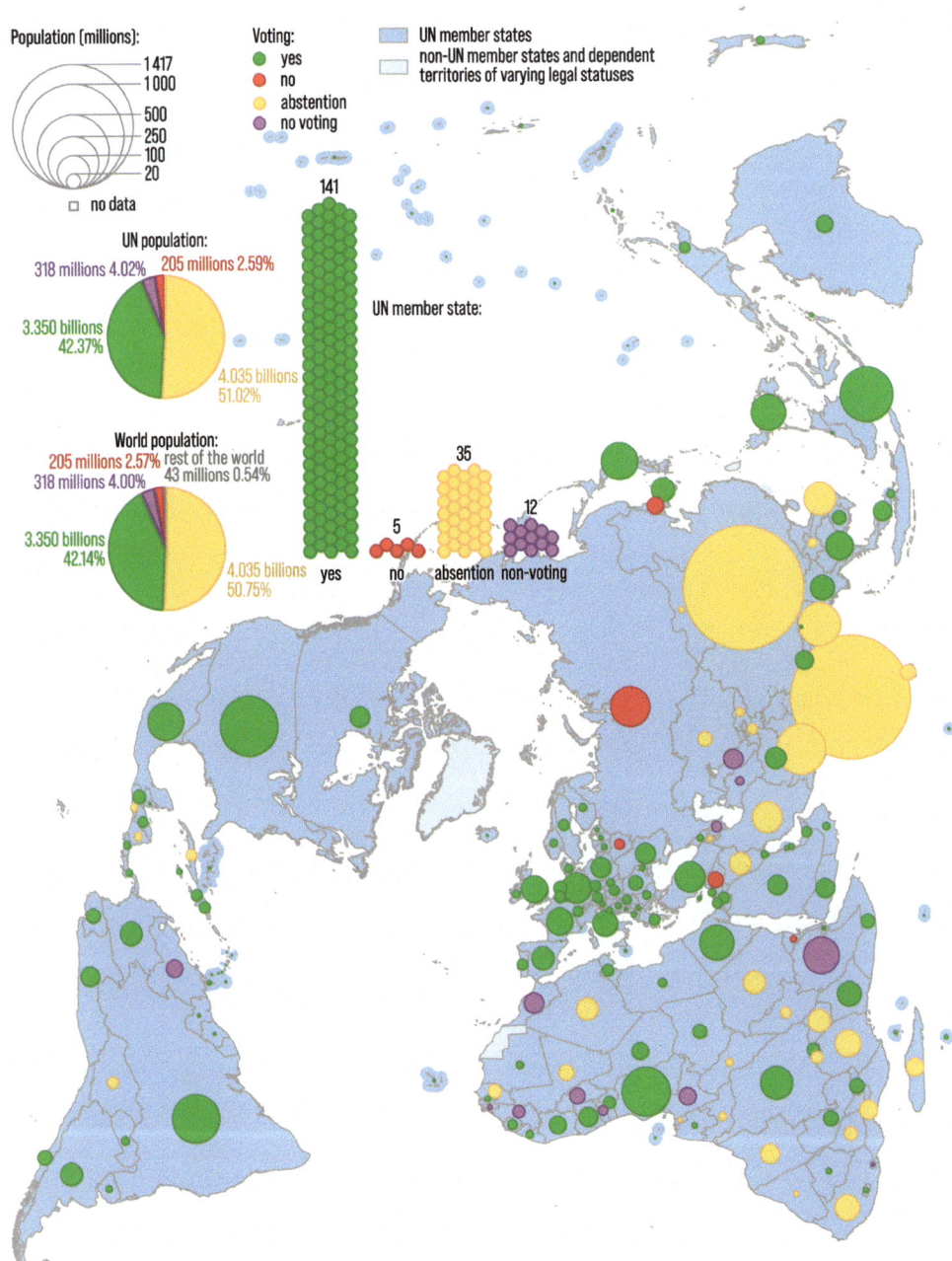

Population (millions):
— 1 417
— 1 000
— 500
— 250
— 100
— 20
□ no data

Voting:
● yes
● no
● abstention
● no voting

UN member states
non-UN member states and dependent territories of varying legal statuses

141

UN population:
318 millions 4.02%   205 millions 2.59%
3.350 billions 42.37%
4.035 billions 51.02%

UN member state:

World population:
205 millions 2.57%   rest of the world 43 millions 0.54%
318 millions 4.00%
3.350 billions 42.14%
4.035 billions 50.75%

yes    no    abstention    non-voting
5    35    12

*Source:* Own elaboration based on United Nations digital library (https://digitallibrary.un.org); The World Bank data (https://databank.worldbank.org)

*Figure 11.15*  2022: UN resolution regarding Russian aggression against Ukraine (second episode) (A/RES/ES-11/1). The UN General Assembly condemns the Russian aggression and calls for withdrawal of its troops. Voting according to UN and world population.

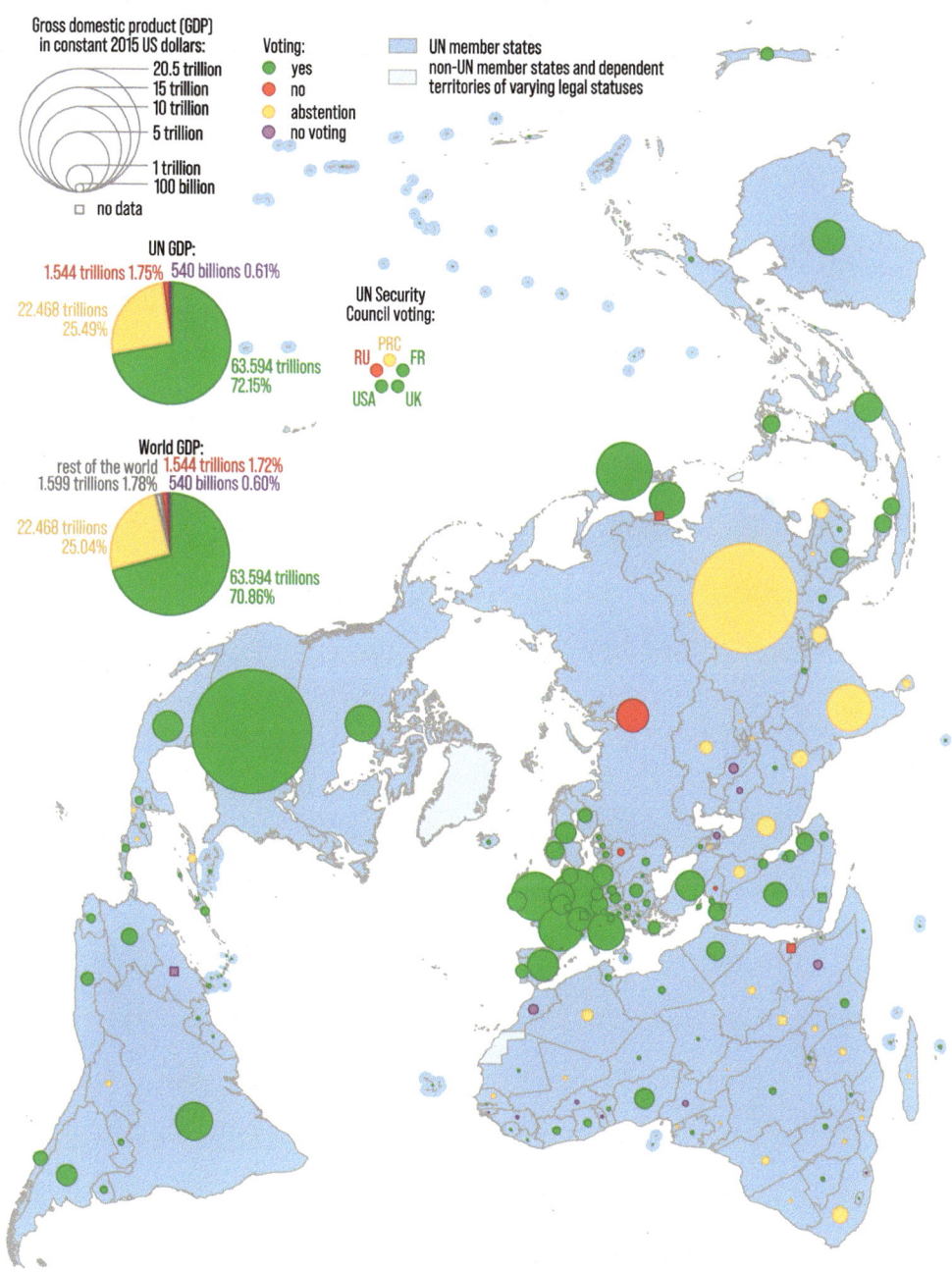

*Figure 11.16* 2022: UN resolution regarding Russian aggression against Ukraine (second episode) (A/RES/ES-11/1). The UN General Assembly condemns the Russian aggression and calls for withdrawal of its troops. Voting according to UN and world GDP.

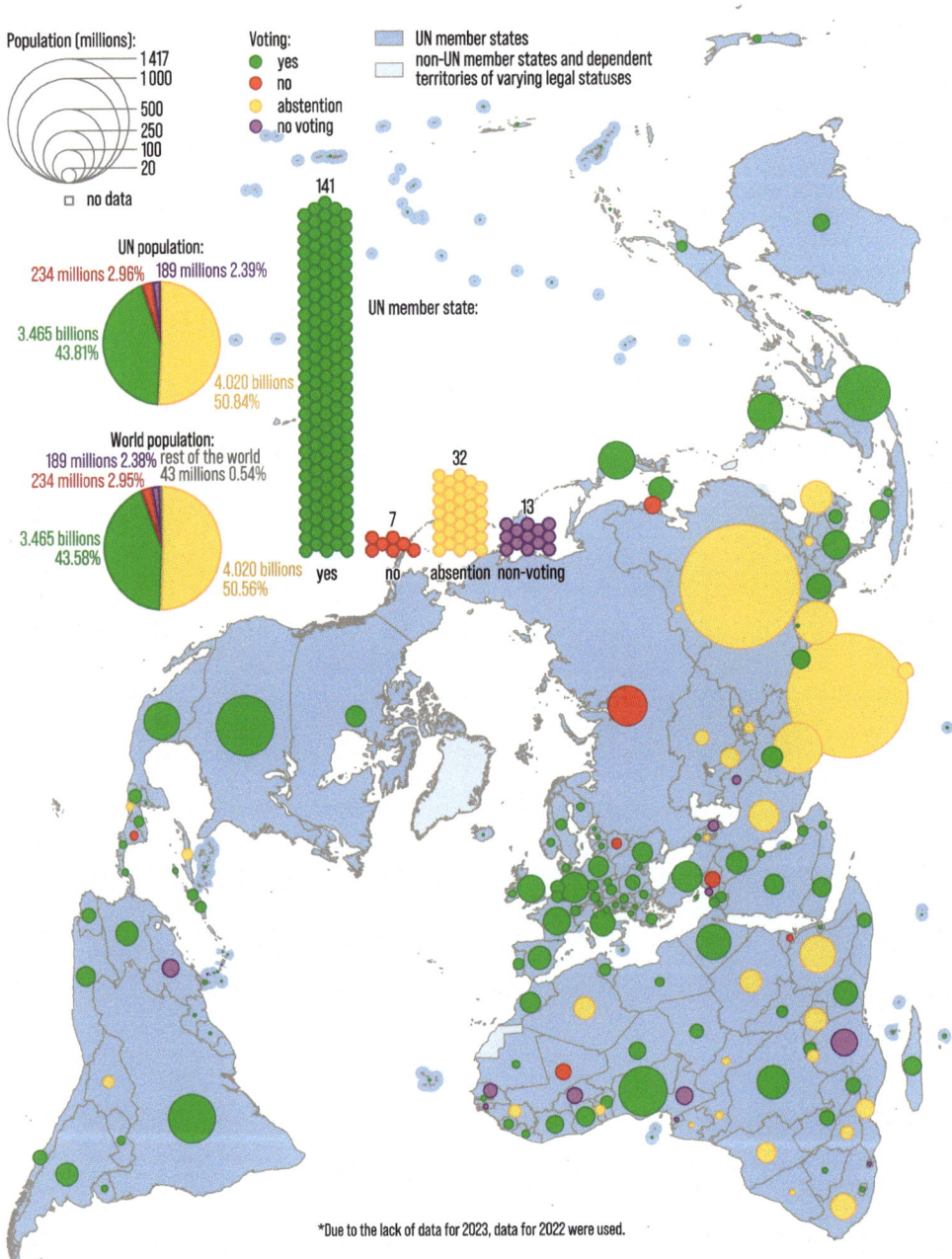

*Figure 11.17* 2023: UN resolution regarding Russian aggression against Ukraine (second episode) (A/RES/ES-11/6). The UN General Assembly reiterates its demand that Russia withdraw its military forces from Ukraine and calls for a cessation of hostilities. Voting according to UN and world population.

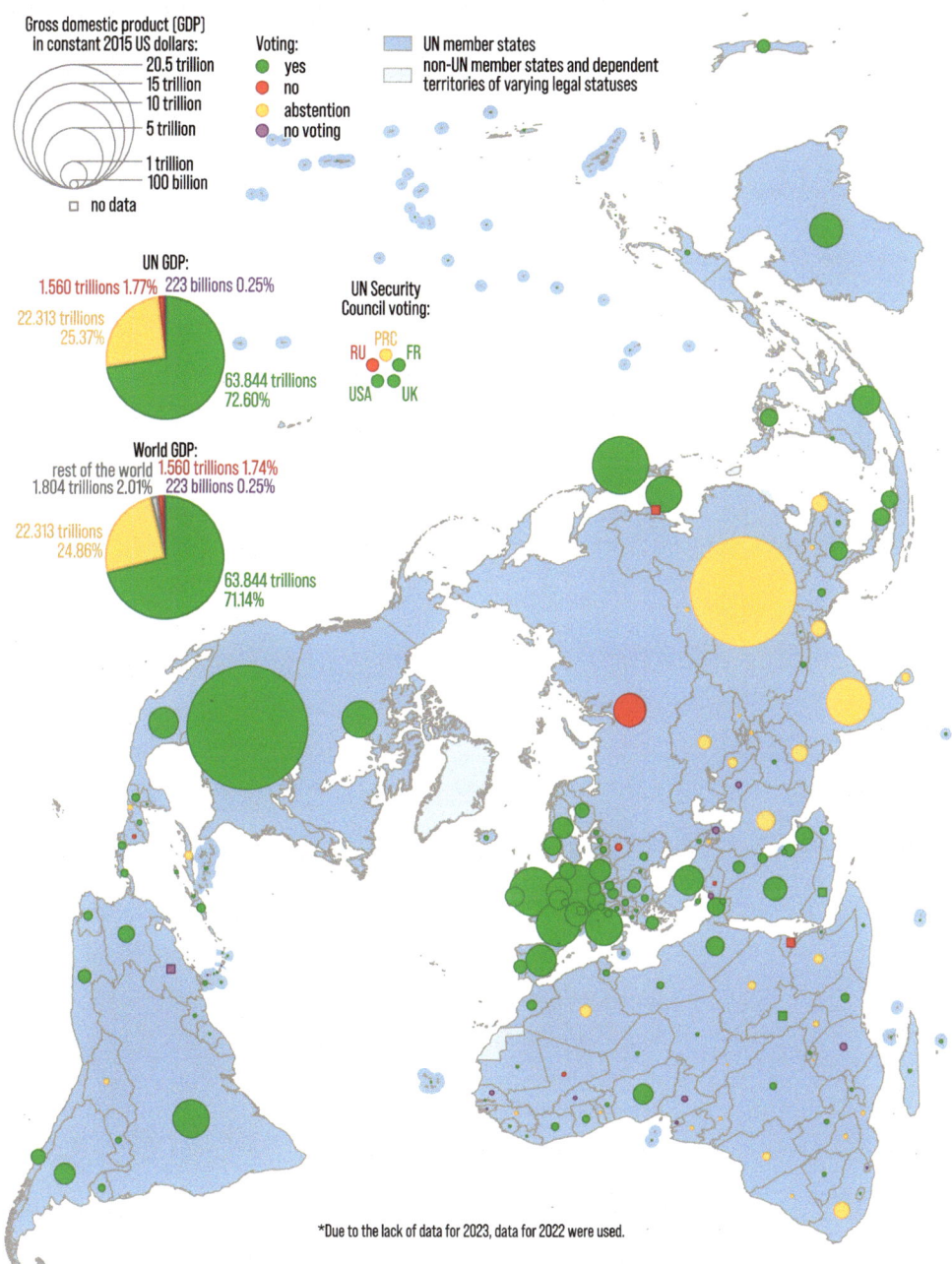

*Figure 11.18* 2023: UN resolution regarding Russian aggression against Ukraine (second episode) (A/RES/ES-11/6). The UN General Assembly reiterates its demand that Russia withdraw its military forces from Ukraine and calls for a cessation of hostilities. Voting according to UN and world GDP.

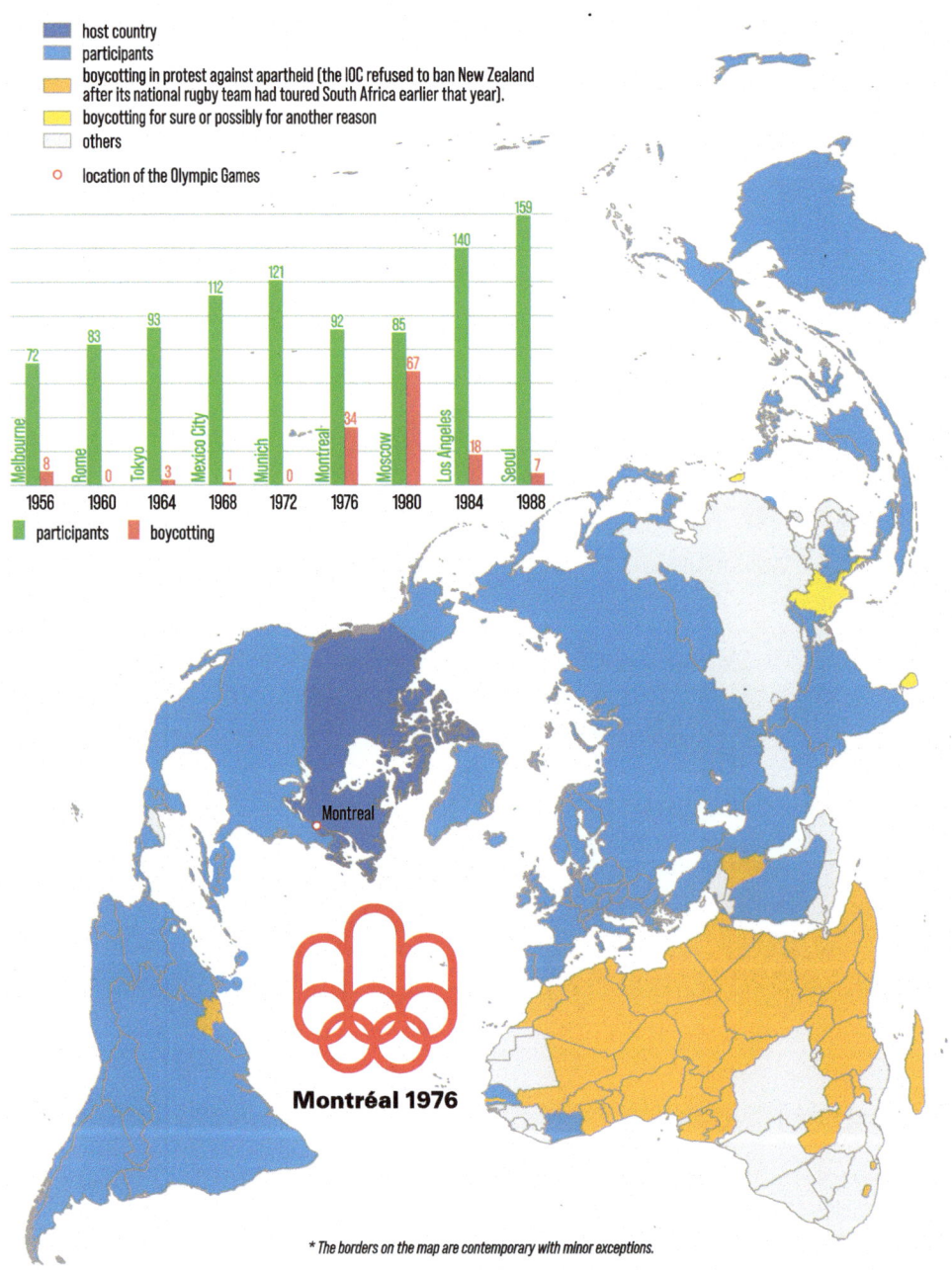

host country
participants
boycotting in protest against apartheid (the IOC refused to ban New Zealand after its national rugby team had toured South Africa earlier that year).
boycotting for sure or possibly for another reason
others

○   location of the Olympic Games

| 1956 | 1960 | 1964 | 1968 | 1972 | 1976 | 1980 | 1984 | 1988 |
| Melbourne | Rome | Tokyo | Mexico City | Munich | Montreal | Moscow | Los Angeles | Seoul |
| 72 | 83 | 93 | 112 | 121 | 92 | 85 | 140 | 159 |
| 8 | 0 | 3 | 1 | 0 | 34 | 67 | 18 | 7 |

■ participants    ■ boycotting

Montreal

Montréal 1976

*The borders on the map are contemporary with minor exceptions.*

*Source:* Own elaboration based on: Keesing's Contemporary Archives; en.wikipedia.org, access 15.09.2023

*Figure 11.19* Boycott of the 1976 Olympics in Montreal, Quebec, Canada, over the participation of New Zealand, which had defied the UN's call for a sporting embargo of South Africa to protest its policy of apartheid

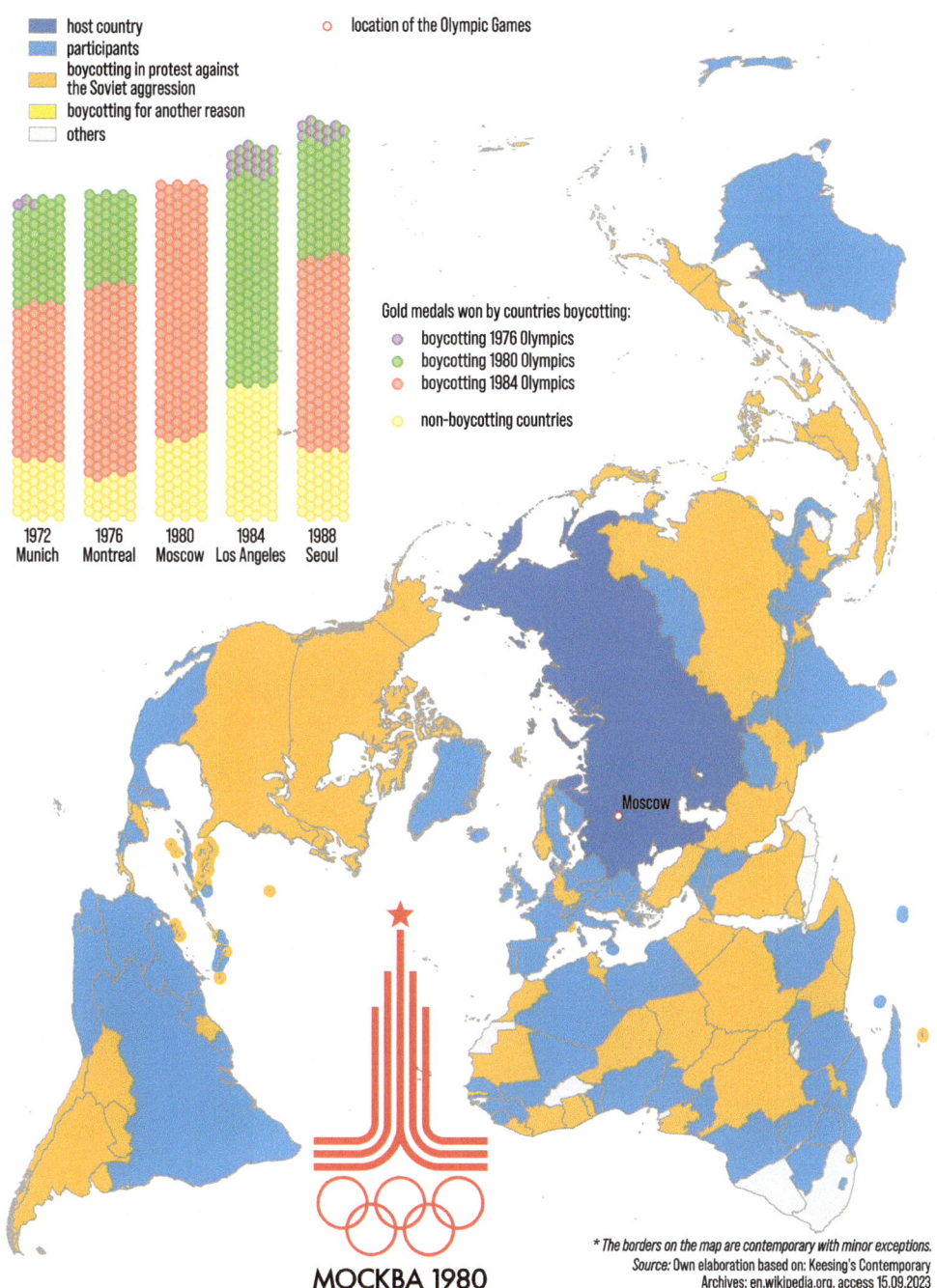

Figure 11.20 Boycott of the 1980 Olympics in Moscow, USSR, to protest the Soviet invasion of Afghanistan

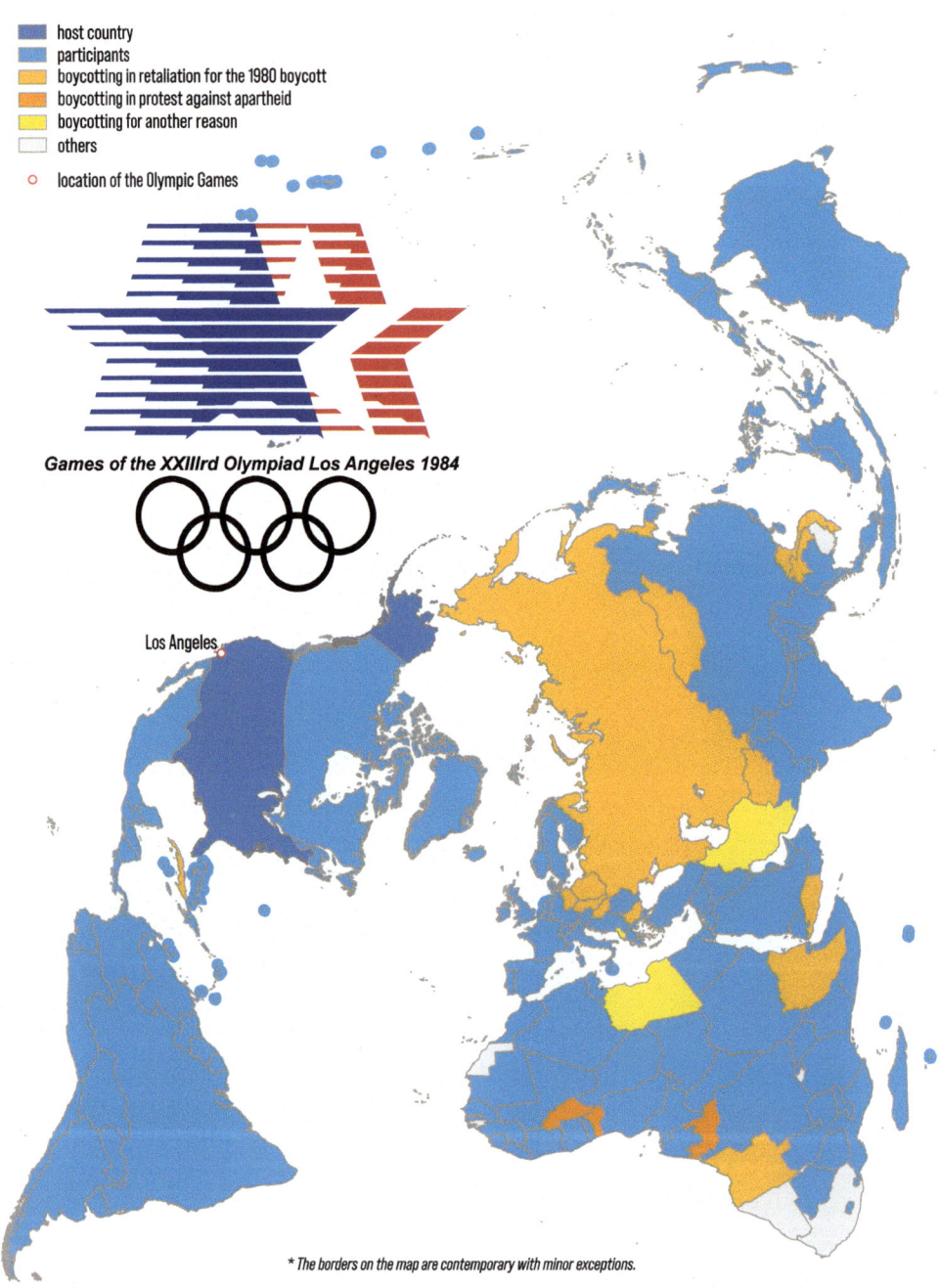

*Figure 11.21* Boycott of the 1984 Olympics in Los Angeles, California, USA, to retaliate against the 1980 boycott

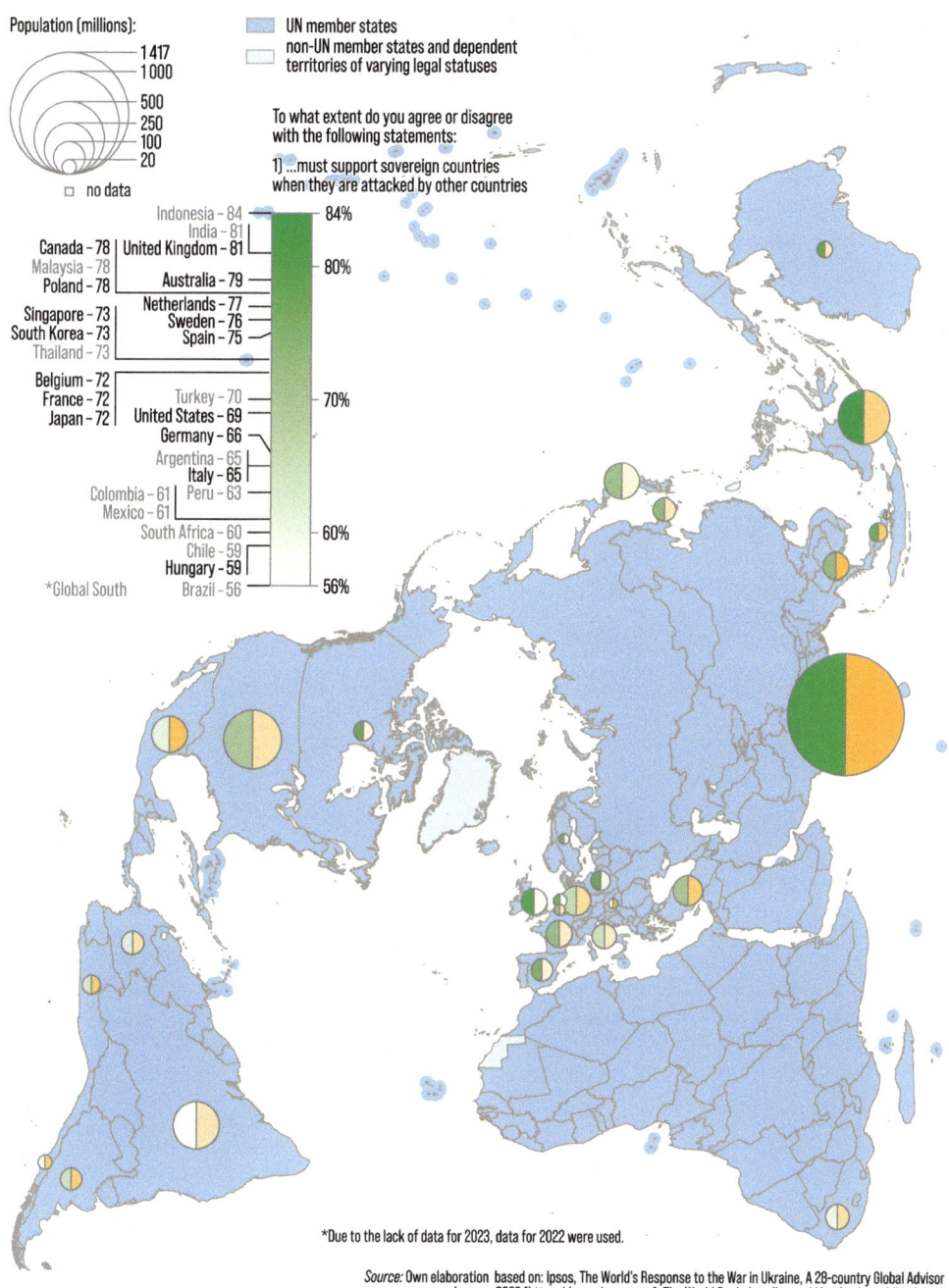

Population (millions):
1 417
1 000
500
250
100
20
□ no data

UN member states
non-UN member states and dependent territories of varying legal statuses

To what extent do you agree or disagree with the following statements:

1) ...must support sovereign countries when they are attacked by other countries

Indonesia – 84 ——— 84%
India – 81
Canada – 78 | United Kingdom – 81
Malaysia – 78
Poland – 78 | Australia – 79 ——— 80%
Netherlands – 77
Singapore – 73 | Sweden – 76
South Korea – 73 | Spain – 75
Thailand – 73
Belgium – 72
France – 72 | Turkey – 70 ——— 70%
Japan – 72 | United States – 69
Germany – 66
Argentina – 65
Italy – 65
Colombia – 61 | Peru – 63
Mexico – 61
South Africa – 60 ——— 60%
Chile – 59
Hungary – 59
*Global South Brazil – 56 ——— 56%

*Due to the lack of data for 2023, data for 2022 were used.

Source: Own elaboration based on: Ipsos, The World's Response to the War in Ukraine, A 28-country Global Advisor survey, January 2023 [https://www.ipsos.com]; The World Bank data (https://databank.worldbank.org)

*Figure 11.22* The international response to the war in Ukraine (January 2023). Opinion of selected societies by population.

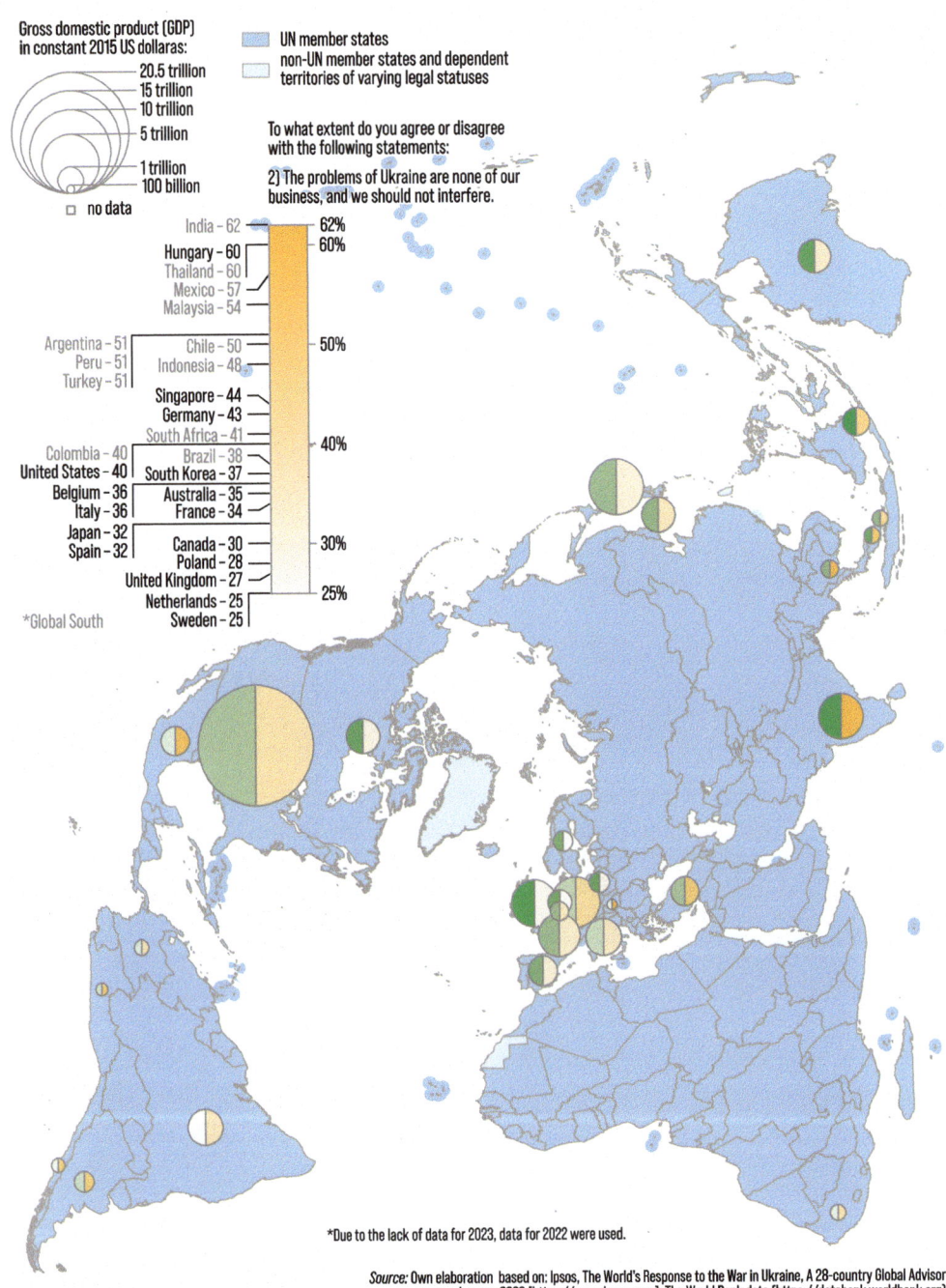

*Figure 11.23* The international response to the war in Ukraine (January 2023). Opinion of selected societies by GDP.

# Conclusions

History often focuses exclusively on one of the chronological realities mentioned at the beginning of this atlas, viz. the history of events, the history of conjuncture, or the history of the *longue durée* (Braudel, 1971). We are naturally drawn to the history of events, as we are immersed in the fast-moving current of history (Kapuściński, 2007) that confronts us every day and captivates us by virtue of its closeness and palpability. Not surprisingly, worldwide attention has been riveted on the Russian invasion of Ukraine on 24 February 2022, although geographical distance from the theatre of war is also significant. Interest and involvement wane as successive 'rings' radiate outward from CEE to the distant Global South. However, when history is examined by considering periods that utilise broad timespans (Braudel, 1971), despite the fact that 'in most cases, history is no more than a catalogue of opening nights. [. . .] "the same thing" is never "the same thing" and a new occurrence of a practically identically situation can produce different results' (Mieroszewski, 1974, p. 3), it enables repetitive 'certain situational arrangements and patterns' to be uncovered (Mieroszewski, 1974, p. 3). They not only allow us to better understand the present buried underneath the noise and informational chaos of the history of events but also allow us to learn about the future in order to avoid bad past scenarios. Observing the 1914–2024 history accurately and listening to its reverberations as much as possible in the perspective of broad timespans suggests that the 1914 July Crisis in fact set in motion a sequence of events that spanned over 100 years, conjoining World War I, World War II, the Cold War, and the Russo-Ukrainian war into one historical process.

DOI: 10.4324/9781003406440-12

# Bibliography

Aristotle (translated by J. A. K. Thomson) (2004) *The Nicomachean ethics*. London: Penguin Books.

Barbag, J. (1978) *Geografia polityczna ogólna*. Warszawa: Państwowe Wydawnictwo Naukowe.

Baugh, D. (2021) *Wojna siedmioletnia. Konflikt globalny (1754-1763)*. Oświęcim: Wydawnictwo Napoleon V.

Bauman, Z. (2004) *Europe. An unfinished adventure*. Cambridge: Polity Press.

Blacksell, M. (2006) *Political geography*. Abingdon: Routledge.

Braudel, F. (1971) *Historia i trwanie*. Warszawa: Czytelnik.

Davies, N. (1996, Polish ed. 1998) *Europe: a history*. Oxford: Oxford University Press.

Dziurok, A. *et al.* (2010) *Od niepodległości do niepodległości. Historia Polski 1918–1989*. Warszawa: Instytut Pamięci Narodowej, Komisja Ścigania Zbrodni przeciwko Narodowi Polskiemu.

Ferguson, N. (2021, Polish ed. 2022) *Doom. The politics of catastrophe*. Dublin: Penguin Books.

Ferro, M. (1997) *Historia kolonizacji*. Warszawa: Oficyna Wydawnicza Volumen, Dom Wydawniczy Bellona.

Fukuyama, F. (1987) 'Soviet strategy in the third world', in Korbonski, A. and Fukuyama, F. (eds.) *The soviet union and the third world*. Ithaca, London: Cornell University Press.

Fukuyama, F. (2001) 'The west has won', *The Guardian (Online)*, 11 October. Available at: http://www.guardian.co.uk/world/2001/oct/11/afghanistan.terrorism30 (Accessed: 7 July 2023).

Furet, F. (1996) *Przeszłość pewnego złudzenia. Esej o idei komunistycznej w XX w.* Warszawa: Oficyna Wydawnicza Volumen.

Gawron, K. (2005) 'Stosunki polsko-czechosłowackie w latach 1918–1939 jako przyczynek do badań nad konfederacją polsko-czechosłowacką 1939–1943', *Historia i Polityka [online]*, 1, 47–78.

Goldin, I. and Muggah, R. (2020, Polish ed. 2022) *Terra incognita: 100 maps to survive the next 100 years*. London: Century.

Hobsbawm, E. (1994, Polish ed. 1999) *The age of extremes: the short twentieth century, 1914–1991*. London: Michael Joseph.

Houellebecq, M. (2010) *La carte et le territoire*. Paris: Flammarion.

Kagan, R. (2003) *Paradise and power: America and Europe in the new world order*. London: Atlantic Books.

Kanet, R. E. (1989) 'The evolution of soviet policy toward the developing world from Stalin to Brezhnev', in Kolodziej, E. A. and Kanet, R. E. (eds.) *The limits of soviet power in the developing world*. Baltimore: The Johns Hopkins University Press.

Kapuściński, R. (2007) *Rwący nurt historii. Zapiski o XX i XXI wieku*. Kraków: Wydawnictwo Znak.

Kelley, R. D. G. (1999) 'A poetics of anticolonialism', *Monthly Review (Online)*, 6 November. Available at: http://monthlyreview.org/1999/11/01/a-poetics-of-anticolonialism (Accessed: 22 October 2012).

Kennedy, P. (1995) *Mocarstwa świata. Narodziny, rozkwit, upadek. Przemiany gospodarcze i konflikty zbrojne w latach 1500–2000*. Warszawa: Książka i Wiedza.

Krajewski, A. (2022) 'Rosyjskie imperium u progu trzeciego rozpadu', *Warsaw Enterprise Institute (Online)*. Available at: https://wei.org.pl (Accessed: 19 June 2023).

Kubiczek, F. *et al.* (eds.) (2003) *Historia Polski w liczbach. Państwo i społeczeństwo. Tom I.* Warszawa: Główny Urząd Statystyczny.

Kuźniar, R. (2005) *Polityka i siła. Studia strategiczne – zarys problematyki.* Warszawa: Wydawnictwo Naukowe SCHOLAR.

Kuźniar, R. (2009) *Poland's foreign policy after 1989.* Warsaw: Wydawnictwo Naukowe SCHOLAR.

Landes, D. S. (1998) *The wealth and poverty of nations.* New York, London: WW Norton & Company.

Maddison, A., (2006) *The World Economy.* Paris: OECD.

Magocsi, P. R. (2019) *Historical atlas of central Europe: third revised and expanded edition.* Toronto, Buffalo, London: University of Toronto Press.

Menkiszak, M. (2021) 'Rosyjski szantaż wobec Zachodu', *Ośrodek Studiów Wschodnich (Online).* Available at: https://www.osw.waw.pl (Accessed: 20 December 2021).

Mieroszewski, J. (1974) 'Rosyjski "kompleks polski" i obszar ULB', *Kultura*, 9(324), pp. 3–14.

Milewski, J. J. (2004) 'Dlaczego Europa? Źródła przyspieszenia wzrostu gospodarczego Europy - przegląd literatury', in Koźmiński, M. (ed.) *Cywilizacja europejska - wykłady i eseje.* Warszawa: Wydawnictwo Naukowe SCHOLAR, Collegium Civitas Press.

Nouschi, M. (2003) *Petit atlas historique du 20e siècle.* Paris: Armand Colin.

Piskorski, J. M. (2001) 'Średniowieczna kolonizacja niemiecka oraz tzw. prawo niemieckie w ujęciu porównawczym', in Samsonowicz, H. (ed.) *Rozkwit średniowiecznej Europy.* Warszawa: Dom Wydawniczy Bellona.

Pol, W. (1878) *Dzieła Wincentego Pola wierszem i prozą. Tom Dziesiąty. Dzieła prozą Wincentego Pola. Tom V. Pisma pomniejsze.* Lwów: F. H. Richter.

Solarz, M. W. (2009) *Północ-Południe. Krytyczna analiza podziału świata na kraje wysoko i słabo rozwinięte.* Warszawa: Wydawnictwa Uniwersytetu Warszawskiego.

Solarz, M. W. (2012) 'Północ-Południe. Anatomia globalnego podziału świata', in Czerny, M. (ed.) *Bieda i bogactwo we współczesnym świecie. Studia z geografii rozwoju.* Warszawa: Wydawnictwa Uniwersytetu Warszawskiego.

Solarz, M. W. (2014a) 'Polska – kilka uwag na temat rozwoju granic, spójności terytorium i zmian położenia geopolitycznego w latach 1667–2014', *Acta Universitatis Lodziensis. Folia Geographica Socio-oeconomica*, 17, pp. 161–184.

Solarz, M. W. (2014b) *The language of global development: a misleading geography.* Abingdon, New York: Routledge.

Solarz, M. W. (2020) 'The language of a globalized world: naming the present day and its worlds', in Brunn, S. D. and Kehrein, R. (eds.) *Handbook of the changing world language map.* Cham: Springer.

Solarz, M. W. (ed.) (2022) *Geograficzno-polityczny atlas Polski. Polska w świecie współczesnym: Perspektywa 2022. Atlas of Poland's political geography. Poland in the modern world: 2022 Perspective.* Warszawa: Biuro Programu NIEPODLEGŁA.

World Bank. data.worldbank.org.

Zabielski, K. (1994) *Finanse międzynarodowe.* Warszawa: Wydawnictwo Naukowe PWN.

Zdziechowski, M. (1993) *Wybór pism.* Kraków: Wydawnictwo Znak.

# Index

Note: page numbers in *italics* indicate a figure on the corresponding page.